The Language *of* Genetics

TEMPLETON SCIENCE AND RELIGION SERIES

In our fast-paced and high-tech era, when visual information seems so dominant, the need for short and compelling books has increased. This conciseness and convenience is the goal of the Templeton Science and Religion Series. We have commissioned scientists in a range of fields to distill their experience and knowledge into a brief tour of their specialties. They are writing for a general audience, readers with interests in the sciences or the humanities, which includes religion and theology. The relationship between science and religion has been likened to four types of doorways. The first two enter a realm of "conflict" or "separation" between these two views of life and the world. The next two doorways, however, open to a world of "interaction" or "harmony" between science and religion. We have asked our authors to enter these latter doorways to judge the possibilities. They begin with their sciences and, in aiming to address religion, return with a wide variety of critical viewpoints. We hope these short books open intellectual doors of every kind to readers of all backgrounds.

Series Editors: J. Wentzel van Huyssteen & Khalil Chamcham
Project Editor: Larry Witham

The Language *of* Genetics

AN INTRODUCTION

Denis R. Alexander

Templeton Press

DARTON · LONGMAN + TODD

Published in the United States of America in 2011 by
Templeton Press
300 Conshohocken State Road, Suite 550
West Conshohocken, PA 19428
www.templetonpress.org

and in Great Britain in 2011 by
Darton, Longman, and Todd
1 Spencer Court
140–142 Wandsworth High Street
London SW18 8JJ
www.dltbooks.com

Library of Congress Cataloging-in-Publication Data

Alexander, Denis R., 1945- author.
The language of genetics : an introduction / Denis R. Alexander.
p. ; cm.
Includes bibliographical references and index.
ISBN-13: 978-1-59947-343-7 (paperback : alk. paper)
ISBN-10: 1-59947-343-7 (paperback : alk. paper) 1. Genetics.
I. Title.
[DNLM: 1. Genetics. 2. Evolution, Molecular. QU 500]
QH431.A378 2011
576.5—dc22

2010051916

Templeton Press ISBN 978-1-59947-343-7

A catalogue record for this book is available from the British Library

DLT ISBN 978-0-232-52878-7

Printed in the United States of America

11 12 13 14 15 16 10 9 8 7 6 5 4 3 2 1

Contents

Preface

As I learned in the summer of 1982, our modern curiosity about genetics is hard to quench, even in the midst of war. That year I was working in Beirut, Lebanon, in the National Unit of Human Genetics in the American University Hospital. Quite suddenly Israeli forces invaded the country, causing much of the population to flee and my research assistant to head for home.

At the time it so happened that a patient was being examined in the hospital suffering from unexplained pains in the back and joints. A bottle of urine, a nice clear yellow, had been passed on to us for examination. It looked normal and was left in the fridge for later examination as the tanks rolled down the highway toward Beirut over that weekend.

By the time I made it back to the lab a couple of days later, something very odd had happened: the urine was pitch-black. I had never seen anything like it before. Memories of my final exams in biochemistry at Oxford came back to me: urine gradually turning black when exposed to oxygen. This must be from a patient with alcaptonuria ("black urine disease"), one of the very first inborn errors of metabolism ever described, a classic but very rare disease often used in undergraduate textbooks to illustrate the impact of defective genes.

The black color results from accumulation of an acid in the body that is then excreted in huge amounts into the urine. I spent the next couple of days distracting myself from the gathering war clouds outside, happily setting up a biochemical test to check that the black color in the urine was indeed this unusual acid. It was. By the time the Israeli tanks had reached the outskirts of Beirut, I had diagnosed our first case of alcaptonuria, a genetic disease that affects only around one person in every million in the United States.

My diagnosis of alcaptonuria was certainly not the first in the Arab world, as I learned later. Just five years before, a study had been published suggesting that an Egyptian mummy dating from 1500 BCE had probably suffered from the disease.[1] Genetic diseases, like names, often have long histories, transmitted down through the centuries. And the marriage of first cousins, so prevalent in ancient Egyptian society, probably made the disease more common.

Part of the romance of genetics is that it impacts history, culture, health, disease, and the trajectories of our own lives, throwing up questions in the process about human identity, uniqueness, moral responsibility, and the extent to which we are willing to go in modifying our environments and our own bodies.

The idea behind the field of genetics is that a few simple biological rules help us to understand the language of life, giving us a grammar to comprehend how living things are built, where they came from, and how to address the reality of disease and our own mortality. Genetics also allows us to understand our human diversity—the color of our eyes, our height, features, and abilities. Regular headlines may make bold claims about the "smart gene" or the "happy gene," but a major goal of this book is to show the remarkable complexity of how genes relate to everything else, especially our environment. If some books suggest that genes control our lives, I try to show that contemporary genetics actually points in the opposite direction, highlighting our human uniqueness and freedom.

This book is not a textbook and makes no claim to be comprehensive. It assumes no background in science, least of all genetics, and aims to build up the knowledge base gradually as the chapters proceed. Technical terms, kept to a minimum, are described upon first introduction. The field is developing incredibly rapidly, so inevitably some of the information will already need updating by the time the book appears. The intention is that, by the end, the language of genetics will no longer appear incomprehensible nor threatening, but rather friendly and familiar. The aim is that the language learned will be sufficient to read the next media pronouncement about a "genetic breakthrough" with informed interest, providing also the confidence to pick up a genetics textbook later in the sure knowledge that much of it will now be comprehensible. The hope is that the book will make people more skeptical: wary of inflated claims, skeptical of results that seem too good to be true, ready to discern the difference between science and ideology.

What is a gene? That might seem like an obvious question, but the answer is not so straightforward. A thread running through this book is the way that our DNA encodes many different kinds of information in a whole variety of ways. Once we begin to see how this information is translated into the building of bodies, we can then appreciate how genetic variation provides the novelty and creativity that drives evolutionary change. The same genetic variation contributes to thousands of different diseases, though at last there are some glimmers of hope that new genetic therapies are beginning to make a difference for at least some of the sufferers.

The book also tackles the trendy new field of epigenetics, the different ways in which the "hardware" provided by the DNA can be supplemented by the "software" of DNA modifications, changing the way in which DNA is "read" by cells, so that the information can even be transmitted down the generations. We see why we need to care for our genomes: our lifestyles can harm them in ways that may be passed on to our grandchildren. Disease arises from epigenetics as much as it does from variation in the DNA itself.

The final chapter goes beyond science to consider some of the big questions that genetics raises: What are its implications for notions of human value and uniqueness? Is evolution consistent with religious belief? Does genetics subvert the idea that life has some ultimate meaning and purpose?

We end the book where we began, with another narrative from Beirut that also goes beyond genetics. However important the language of genetics may be, it is never the only story to be told, and will never have the final word.

Acknowledgments

I WOULD LIKE to thank all those who have contributed so helpfully to the preparation of this book. My special thanks goes to Dr. Nicole Maturen, who carried out some of the background research, helped bring nuance and insight into many sections, and was also responsible for organizing all the figures. My thanks are also due to Andrew Short and Sharon Capon for their assistance in the preparation of figures and to Dr. Graeme Finlay for providing Figures 7.3, 7.4, 7.5, and 7.6. I am also very grateful to all those kind experts from different fields within the biological sciences who read an earlier version of the manuscript either in whole or in part and made many helpful corrections and suggestions: Dr. Ruth Bancewicz (St. Edmund's College, Cambridge), Dr. Graeme Finlay (University of Auckland, New Zealand), Prof. Keith Fox (Southampton University, U.K.), Prof. Jeffrey Hardin (University of Wisconsin–Madison, U.S.), Sir Brian Heap, FRS (Cambridge University, U.K.), and Dr. Cara Wall-Scheffler (Seattle Pacific University, U.S.). Any errors that remain are my responsibility alone.

The book is dedicated to all those members of the National Unit of Human Genetics at the American University of Beirut Medical Center who pursued their mission with such dedication during the difficult years of 1980–1986. They will not be forgotten.

The Language *of* Genetics

CHAPTER 1

The Birth of Genetics

THE BIRTH of scientific ideas is never straightforward. Theories twist and turn. Blind alleys are pursued. The different results that seem so obviously connected with the benefit of hindsight are at first kept in splendid isolation, or even in opposition.

The birth of genetics—the word itself derived from the Greek "to give birth"—illustrates all these complexities. This is no heroic tale of one triumphant discovery after another, but a story about a dark maze full of groping investigators whose insights gradually enabled us to understand the language of genetics, a research field that now continues to expand its range of discoveries at a breathless pace.

Our own curiosity about what we now call "heredity" and "genetics" is not as modern as it may at first seem. Four thousand years ago the Assyrians and Babylonians were manipulating genes when they pollinated date palms, not knowing a thing about genetics. An Assyrian bas-relief sculpture shows the artificial pollination of date palms at the time of King Ashurnasirpal II of Assyria, who reigned from 884 to 859 BCE.[1] The foundation we stand on today is built on a rich history, which began with questions about inheritance of traits in humans, animals, and plants. In time, the search was on for the pattern of this inheritance, and finally—in the twentieth century—the mechanism that eventually provided the basis for genetics. The birth of genetics is a story that takes us from ancient Greek speculation about twins, through the key discoveries of Gregor Mendel in the nineteenth century, and on to the 1953 discovery by James Watson and Francis Crick of the structure of DNA, the double-helix, which has become the biological icon of our age.

Early Ideas about Heredity

The early Greek philosophers speculated extensively about the mysteries of human heredity. Hippocrates (c. 460–370 BCE), considered the father of modern medicine, expounded an idea that much later came to be known as pangenesis, in which the material of inheritance is collected from throughout the body, delivered to the reproductive organs, and passed to the embryo at the moment of conception. As Hippocrates wrote, "The offspring resembles its parent because the particles of the semen come from every part of the body."[2]

Aristotle (384–322 BCE) opposed this idea of inheritance by reassembled particles, objecting, not unreasonably, "How could there be such particles for abstract characters as voice or temperament, or from such nongenerating sources as nails or hair?"[3] Instead Aristotle saw inheritance in more qualitative terms in which sperm provided the "active element," bringing the offspring to life, whereas the female contributed the nutrition that would help the offspring to grow. Aristotle considered two types of explanation for development. In the preformationist idea, a miniature individual exists in the egg or sperm, and then begins to grow into the offspring upon stimulation. In the theory of epigenesis, which Aristotle himself favored, the new organism develops from an undifferentiated mass by the addition of parts. As happens so often in the history of ideas, the Greek philosophers thus set the general agenda for the discussion about inheritance for the next two thousand years. Was it a question of physical particulate inheritance; or the passing on of preformed miniature individuals, like preformed Russian dolls, one inside the other; or the development of new organisms out of an undifferentiated mass? All these suggestions played important roles in the discussions that followed over the centuries.

Accurate observations of familial inheritance were more common than satisfying explanations of how the pattern of inheritance worked. Rules to prevent the consequences of what we now call hemophilia, in which blood fails to clot properly, can be found in the Jewish Talmud, and in 200 CE Rabbi Judah the Patriarch exempted a third son from circumcision if two elder brothers had bled to death.[4] Even more striking is the Talmudic exemption that was also provided to the boy's male cousins, providing that they were sons of his mother's sisters, but not sons of his mother's brothers or his father's siblings. This exemption recognizes what we now call an X-linked pattern of inheritance, which is explained below.

Identical twins also drew much early attention and speculation. St. Augustine (354–430) argued against astrology on the grounds that twins born at virtually the same time, under the same planets, could have very different personalities. Much later Martin Luther (1483–1546) echoed the same argument, ridiculing astrology by pointing out that Esau and Jacob in the Old Testament were twins, yet had very different characters.

As the experimental method gained broader application with the scientific revolution of the sixteenth and seventeenth centuries, so some key findings were made that helped to lay the groundwork for the later science of genetics. James I's personal physician, William Harvey (1578–1657), who described the heart as a pump and explained its role in the circulation of the blood, also published *On the Generation of Animals* in 1651, arguing that all living organisms arose from eggs. Such was the fascination with the very small, aroused by new discoveries with the microscope, that preformationist ideas gained more attention, with either eggs or sperm being touted as the location of the "homunculus," the preformed miniature individual destined to become the new offspring. But as critics pointed out, the theory did not explain why offspring were such a mingling of the features of both parents.

The Swedish botanist Carl Linnaeus (1707–1778) first published his famous system of classification of all known living things in 1735, a classification that provides the basis for all further classifications right up to the present day. At the time Linnaeus believed that the number of species had been fixed at the time of the creation. But as Vítězslav Orel comments, "Subsequent experimental crossing of plants convinced him that hybridization gave rise to combinations of parental traits. He thought the genus rather than the species to be the basic unit of creation, and now admitted the possibility of new species appearing in nature and disappearing from it. He formed an open system, interpreting it in harmony with the Creator's design."[5]

The introduction of the microscope opened up a fascinating new world of detailed biological structure that readily lent itself to mechanical types of description. Using the microscope, the polymath Robert Hooke (1635–1703) described for the first time in his famous work *Micrographia* (1665) the existence in plants of what he called "cells," the name suggested to him by their resemblance to monks' cells. Over a century later, Robert Brown (1773–1858), an extraordinarily gifted microscopist,[6] was the first to identify the existence of the cell nucleus, where we now know the genetic material resides—at least in those cells that contain a nucleus, known as eukaryotes (in prokaryotes, such as bacteria, the genetic material is not separated from the rest of the cell).

By the mid-nineteenth century, botanists had carried out systematic breeding experiments, and the microscope had opened up the world of the cell and its nucleus. But the puzzling complexity of different patterns of inheritance eluded any clear explanation. Charles Darwin (1809–1882), brilliant naturalist that he was, whose theory of natural selection was destined to change the face of biology, nevertheless failed to uncover the laws of inheritance. Had he done so, evolution would have been a more complete theory much earlier.

Darwin set out his own views on inheritance in the second volume of his 1868 work, the *Variation of Plants and Animals under Domestication*. There Darwin presented his theory of pangenesis, apparently unaware how similar his account sounded when compared to that of Hippocrates more than two thousand years earlier:

> I venture to advance the hypothesis of Pangenesis, which implies that every separate part of the whole organisation reproduces itself. So that ovules, spermatozoa, and pollen-grains,—the fertilised egg or seed, as well as buds,—include and consist of a multitude of germs thrown off from each separate part or unit.[7]

Darwin gave the name "gemmules" to these hypothetical physical units that were gathered up from all parts of an organism and "packaged" in some way in the eggs and sperm, from there to be passed on to the offspring. Darwin believed, like his forerunner Jean-Baptiste Lamarck (1744–1829), in the inheritance of acquired characteristics: the external environment could modify the inheritable gemmules. He also maintained that inheritance resulted in a "blending" of the characteristics of both parents, with the "gemmules" playing a key role in the blending process.

But Darwin, always self-critical to a fault, was only too aware that his theory of inheritance was incomplete and certainly did not explain all the data; it was a "provisional hypothesis or speculation," as he modestly concluded. This became apparent through his own plant breeding experiments carried out in the back garden of his home in Kent, Down House, and also by his study of the inheritance of deafness in various families. The varying patterns of inheritance puzzled Darwin. He could find no obvious consistency that lent itself to a clear explanation.

Genes—the Birth of an Idea

Ironically, at the very time that Darwin was puzzling over the inheritance of deafness and proposing his theory of pangenesis, which was in fact wrong, an Augustinian Moravian monk named Gregor

Mendel (1822–1884) had not only carried out the key experiments that would lay the foundation of modern genetics, but also published his results in 1866.

Moravia at the time was, in the words of the historian Edward Larson, "a region of the Austro-Hungarian Empire, then a proto-modern police state with quasi-medieval remnants of ecclesiastical privilege."[8] Mendel was the only son of a peasant farmer. In 1843 he gained admission to the wealthy and scholarly St. Thomas Monastery of the Augustinian Order near the Moravian capital of Brünn (now Brno in the Czech Republic), where he remained for the rest of his life, eventually becoming its abbot. Following ordination as priest in 1847, Mendel was assigned to teach in a secondary school in the city of Znaim (now Znojmo), but he failed the exam necessary to obtain a teaching certificate. He then went to study mathematics and biology in the University of Vienna in 1851, and there he gained the analytical expertise that would be so useful in his later breeding experiments.

Following graduation, Mendel returned to teaching in Brünn, where again he attempted the teachers' certificate exam. This time he withdrew at the last moment due to illness, some think brought on by his own anxiety about taking the exam again. Mendel did continue to teach science part-time in the technical school close to the monastery, and by all accounts was an inspiring teacher. Clearly, passing exams successfully may not necessarily be life's most important hurdle. Indeed, during the following period—1856 to 1863—Mendel carried out the key plant breeding experiments in the monastery gardens that were destined to change the face of modern biology.

As Mendel was careful to point out in the publication of his results, a long history of plant breeding experiments preceded his own. The brilliance of his approach was to link careful experimental observations with the mathematical analysis that revealed what much later came to be known as "Mendel's Laws of Inheritance." His experiments essentially consisted of "growing, crossbreeding, observing, sorting and counting nearly thirty thousand pea plants of various carefully selected varieties."[9] Like much successful work in science, the experiments involved a judicious choice of materials to work with, a lot of patience, a sharp eye for detail, and smart analytical skills. Mendel also had a great love of fine food and good cigars, both reportedly consumed in prodigious quantities, which no doubt helped with the analysis.

Mendel's experiments revealed several key findings. His starting varieties of pea plants had bred true for many generations. Today we would say that they were genetically pure lines, displaying reproducible traits over many generations. This was an important factor in his success. When Mendel cross-hybridized these different varieties, the traits inherited by the next generation of peas (the "hybrids") were "particulate"—their seeds were either wrinkled or smooth, or the plants were either tall or short. The hybrids showed only one of the two possible character traits present in the parents, inconsistent with the idea of "blending inheritance" in which different traits merged with each other. Mendel also noticed that some traits were "dominant" and some were "recessive." When he crossed the tall pea plants with the short pea plants, then the first generation was all tall, but the ratio of tall to short plants in the second generation came to approximately three to one—tall was a dominant trait, and short was a recessive trait. If he crossed tall with tall, then he got only tall, and likewise short with short yielded only short plants. Experiments with peas having multiple different characteristics suggested that each trait—for example, height, color, texture—was inherited independently through subsequent generations.

Mendel read the paper summarizing his results at two meetings of the Brünn Natural History Society in 1865, but for the next thirty-five years his paper, buried away in the little-known *Proceedings of the Natural History Society of Brünn,* was cited only a few times and its importance went unrecognized. There is no evidence, for example, that Darwin knew about Mendel's work. This provides a good illustration of the way science works. New findings not only have to be widely disseminated to gain attention, preferably in a high-profile journal, but they also have to fall on ready soil. Some results can be so well ahead of the research field as a whole that they are simply ignored, and Mendel's key results seem to have suffered such a fate for the rest of the century.

Soon after Mendel's publication, in 1869, Friedrich Miescher discovered a weak acid in the nuclei of white blood cells (isolated from the pus on bandages collected from the local hospital in Tübingen, Germany). It would be nearly a century until that substance, deoxyribonucleic acid (DNA), was identified as the molecule responsible for Mendel's results. In the meantime, the mystery of inheritance continued to pique the curiosity of many.

August Weismann (1834–1914) made the important observation, published in 1893, that the body had two different types of cells, the "somatic cells" that made up the bulk of the body and did not pass on their information to succeeding generations, and the "germ cells" (the egg and sperm cells) that did pass on information. Moreover, he noted that the two types of cell replicated in different ways. Somatic cells came from germ cells, but not vice versa, rendering the inheritance of acquired characteristics impossible. As such, Weismann's finding contradicted the theory of pangenesis. To make quite sure, he chopped the tails off fifteen hundred rats, repeatedly over twenty generations, and reported that no rat was ever born in consequence without a tail. It really did seem that the property of being a tailless rat was not inherited, though to be fair on Darwin's theory of pangenesis, he had always insisted that major changes in an organism, like a mutilation, could not be inherited. Ironically Weismann's own theory of "germinal selection" suggested that some changes in the germ line could be brought about by beneficial changes imposed by the environment that could then be inherited. The Lamarckian idea that acquired characteristics could be inherited turned out to be very persistent.

Finally, at the turn of the century, Mendel's seminal work was rediscovered and extended by three fellow plant breeders: Hugo de Vries (1848–1935) in Amsterdam, son of a Mennonite deacon who later became prime minister of the Netherlands; Carl Correns (1864–1933) in Tübingen, who was encouraged to study botany by a correspondent of Mendel; and Erik von Tschermak (1871–1962) in Ghent, whose grandfather had taught Mendel during his time in Vienna. All three had been using different plant breeding systems to investigate inheritance, and each confirmed a three-to-one ratio between dominant and recessive traits in his own system. With varying degrees of alacrity, they recognized that Mendel's work had foreshadowed their own, and together they helped to launch Mendel to the central place that he still enjoys in the history of genetics.

William Bateson, first professor of genetics at Cambridge University, widely publicized these breakthroughs in his monumental work *Mendel's Principles of Heredity*, in which Bateson spelled out the importance of Gregor Mendel's rediscovered ideas.[10] In the very same year, 1902, Sir Archibald Edward Garrod made the first detailed biochemical description of several inborn errors of metabolism, of which alcaptonuria was one.[11] Garrod commented in his paper on the remarkable number of patients with alcaptonuria whose parents were first cousins, just as they are in Lebanon and many other countries today, using Mendel's results to explain the pattern of heredity. It was Bateson who first coined the word "genetics" (from the Greek *gennō, γεννώ*; "to give birth").[12]

All these results provided striking confirmation of the particulate theory of inheritance. In 1909 the Danish botanist Wilhelm L. Johannsen (1857–1927) introduced the term "gene" to replace older terms like factor, trait, and character: the word was deliberately chosen to contrast with "pangene," the earlier term associated with the now discredited ideas of pangenesis. Johannsen also introduced the terms "genotype" and "phenotype," to refer to the genetic information contained within germ cells and the visible characteristics of an organism, respectively.

The first few decades of the twentieth century were marked by a range of insights into the newly discovered explanatory powers of genetics. A team of researchers at Columbia University led by Thomas Hunt Morgan, the first American biologist to receive the Nobel Prize, pioneered genetic studies using the fruit fly *Drosophila.* The studies met with great success, showing that genes were strung out on chromosomes, as Morgan put it, "like beads on a string." Mutant genes for eye color followed a recessive pattern of inheritance, so Morgan was able to show that Mendelian inheritance applied as much to animals like *Drosophila* as it did to plants. In 1915 Morgan coauthored with three other collaborators the famous book *The Mechanism of Mendelian Inheritance,* which completed a revolution in scientific thought by placing genes at the center of biologists' ideas about heredity.

By 1927 the American geneticist Hermann Joseph Muller (1890–1967) had demonstrated that it was possible to induce mutations in *Drosophila* by means of radiation using X-rays. Many of these mutations were stable over many generations and demonstrated a Mendelian pattern of inheritance.

Investigation of a bread mold called *Neurospora* during their work at Stanford University enabled Edward Tatum and George Beadle to show in 1941 that mutations in genes could cause defects in specific metabolic pathways inside the cells, leading to the idea that one gene encodes one enzyme. Enzymes are proteins that catalyze chemical reactions. Later it was recognized that one gene can encode not just one enzyme, but any kind of protein.

In 1944 DNA was identified as the genetic material. This key finding was carried out at the Rockefeller Institute in New York by Oswald Avery (1877–1955). Avery's father was a Baptist minister who moved his family from Canada when he was appointed pastor of the Mariner's Temple Baptist mission church on New York's Lower East Side. Oswald then spent the rest of his life in the States. Intriguingly, another Baptist minister, Frederick Taylor Gates, had originally persuaded John Rockefeller, himself very involved in New York Baptist church life, to set up the Institute in the first place (in 1901).[13] Avery's research team found that the characteristics of one strain of bacteria could be transferred to another purely through DNA and not via proteins. At the time scientists thought that proteins were the chemicals of inheritance because they seemed to display a huge amount of variety, whereas no one knew how nucleic acids like DNA might convey variation. Avery's results were initially greeted with disbelief, the world being distracted by the turmoil of the Second World War. By the time the significance of Avery's results was fully appreciated, he was dead, and Nobel Prizes cannot be awarded posthumously.

Yet Avery was entirely correct, and his findings laid the groundwork for the era of molecular biology, fully launched in 1953 with the publication of the double-helical structure of DNA by James Watson and Francis Crick. As they mentioned laconically at the end of their famous paper in the journal *Nature,* a paper barely one page in length, "It has not escaped our notice that [the structure] we are postulating immediately suggests a possible copying mechanism for the genetic material."[14] Genes were now no longer the rather vague "inheritable principles" of earlier generations, but specific entities located within the DNA, the very structure of which unlocked the mystery of heredity.

Figure 1.1 provides a summary outline of some of the key findings in the history of genetics that bring us into the twenty-first century. With the idea of the gene now born, we can look more closely at its role in inheritance, its structure, and the fascinating way in which the very meaning of the word continues to twist and turn like the double-helix itself, right up to the present day.

Patterns of Inheritance

Apologies to all gardeners, and to Mendel, but a lot of people find mapping inheritance through the generations using smooth or wrinkled seeds pretty boring. But if we take Mendelian genetics

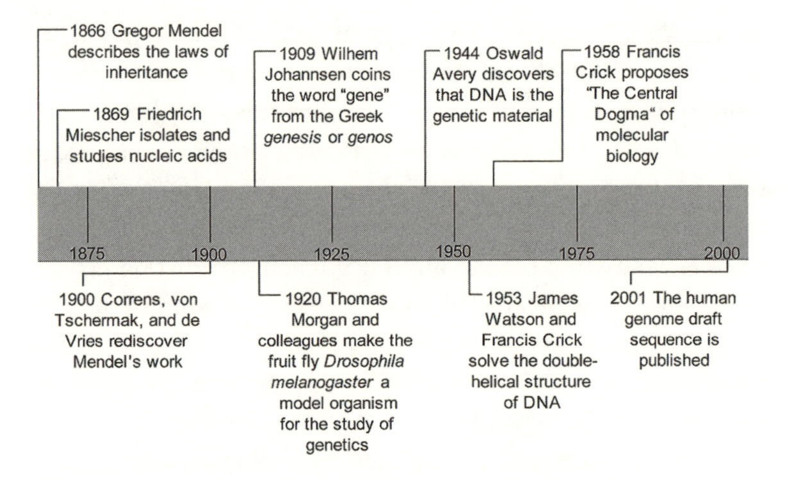

Figure 1.1. Major events in the development of genetics as a scientific discipline, from Mendel's discovery of the laws of inheritance in 1866 to the completion of the draft of the human genome in 2001.

and apply it to our own health, that's a different matter altogether, and because genes work the same way in plants or in animals, exactly the same principles apply to us as they do to the shapes of seeds. Defective genes in humans are the cause of around six thousand distinct single-gene disorders, which occur in about one out of every two hundred births, and more are being discovered all the time.[15]

The inheritance of human genetic diseases follows three principal patterns, which correspond to recessive, dominant, and X-linked genes, and in this section we consider these three mechanisms one at a time. Morgan's idea of genes as beads strung along on a string is all you need to know to track the different patterns. The string in this analogy is the chromosomes, which come in pairs. Different animals and plants have different numbers of chromosomes; humans possess twenty-two pairs of chromosomes, and the twenty-third pair is either XX for females or XY for males, thereby defining our sex differences. The germ cells contain only a single set of unpaired chromosomes, and sperm contain either an

X or a Y chromosome in a roughly 50-50 ratio. Since all eggs contain a single X chromosome, if a sperm with an X fertilizes the egg, then the child is female, but if it's a Y chromosome sperm, then the child is male. Fortunately the swimming abilities of X and Y sperm are pretty similar, which helps to keep the population reasonably balanced. In fact, it's not the first sperm that reaches the egg that wins. Quite a few kamikaze sperm launch themselves on the egg first and weaken its membrane before the lucky one gets inside. So we should all be thankful to those kamikaze sperm that blazed the trail for the sperm that eventually led to each of us.

Inheritance of Recessive Genes

The genetic disease cystic fibrosis provides a clear example of the inheritance of a recessive gene. Cystic fibrosis affects the lungs and digestive systems of about thirty thousand children and adults in the United States and twenty thousand in Europe. The disease occurs mostly in whites whose ancestors came from northern Europe, although it affects all races and ethnic groups. It is the most common recessively inherited disorder in Caucasians (about one in three thousand births).

The defective gene encodes for a protein that functions as a "channel" (meaning a very small hole) in the membranes of cells to allow the passage of chloride ions, commonly found in table salt (sodium chloride). People with cystic fibrosis find that their own skin tastes salty when they lick it. Eighteenth-century German literature warns, "Woe is the child who tastes salty when kissed on the forehead, for it is cursed and soon must die," a curse now thankfully confronted by modern medicine. When the chloride channel is defective, a thick, sticky mucus develops that clogs up the lungs, leading to infections. The mucus also obstructs the normal working of the pancreas so it can no longer release the enzymes properly that are needed to digest food.

Figure 1.2a shows the pedigree of a family in which cystic fibrosis occurs. Boxes represent males and circles females. An empty white symbol represents a family member with two normal copies of the gene that encodes the chloride channel. A half-black symbol represents a family member with one normal gene and one defective gene (a condition known as "heterozygous"). The normal gene compensates for the defective recessive gene, so this person does not suffer from the disease; in fact, he cannot know that he is a carrier of a defective gene before having children unless he has a genetic test. The black symbols represent family members with two copies of the defective gene, and they are the ones with cystic fibrosis (known as "homozygous" for the defective gene). More than 70 percent of U.S. cases are diagnosed by the age of two. There are generally no mistaking the consequences when both copies of the gene are defective.

Note that this family tree, an example from real life, illustrates exactly the principles of inheritance that Mendel worked out in the garden of his monastery by breeding pea plants. Males and females can both carry the defective gene. Because offspring inherit their genes 50-50 from their mother and father, if only one parent is a carrier of a single defective gene, then an average of 50 percent of the couple's offspring will also be carriers (the "heterozygotes"). The 50-50 chance of the defective gene being passed on in the carrier parent's sperm (or egg) applies to each separate birth, and just as tossing a coin can produce three heads in a row, so it's quite possible that in the family shown in Figure 1.2a, all three children in the second generation could be carriers, or none at all. As it happens, two children are carriers and one is not.

What happens when two carriers of the cystic fibrosis gene have children, as seen in the third genereation of the family shown in Figure 1.2a? For each child, the chances are as follows: 25 percent that the child has two copies of the defective gene ("homozygous"), and thus has the disease; 25 percent that the child is not a carrier of a defective gene at all; and 50 percent that the child has a single copy of the defective gene ("heterozygous"). This is exactly what Mendel determined for a recessive trait: the ratio of disease-

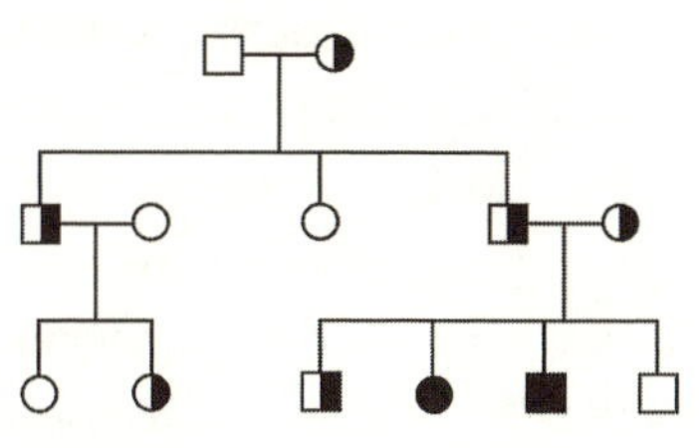

a) Cystic Fibrosis

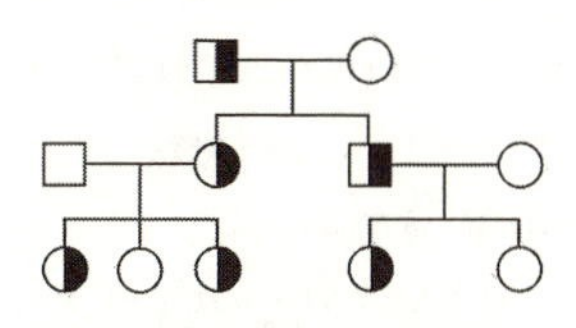

b) Familial Hypercholesterolemia

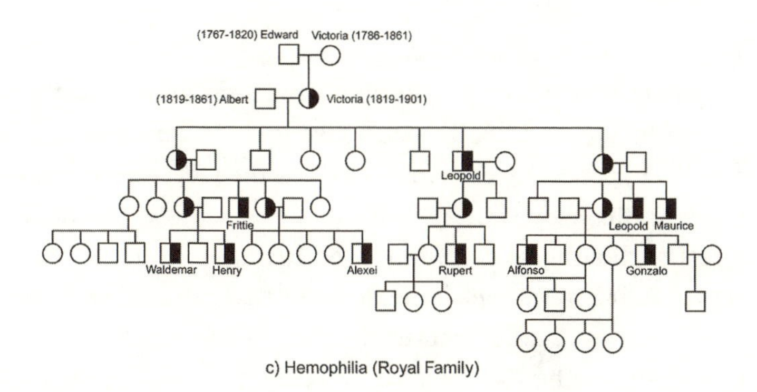

c) Hemophilia (Royal Family)

Figure 1.2. Three pedigrees of families suffering from three diseases, each exhibiting a different mode of inheritance. Circles represent females and squares represent males. Filled shapes indicate that the person carries two copies of the disease gene; half-filled shapes indicate that the person carries one copy of the disease gene; unfilled shapes indicate that the person carries two normal copies of the gene of interest. a) Cystic fibrosis, an example of recessive inheritance. Those with one copy of the disease gene are disease-free and often called "carriers"; those with two copies of the disease gene manifest the disease. b) Familial hypercholesterolemia, an example of dominant inheritance. Those with one copy of the mutant gene manifest the condition. c) Hemophilia, an example of X-linked inheritance. This pedigree shows how the condition was inherited through the royal family of Queen Victoria of Great Britain. It is assumed that the mutation causing hemophilia arose *de novo* in Queen Victoria or in the germ-line cells of one of her parents. Females who carry a copy of the mutant gene do not manifest the disease due to the presence of a healthy copy of the gene on their second X chromosome. Males, however, have only one X chromosome. If that X chromosome bears a mutant gene, the male will develop hemophilia, as did the sons, grandsons, and great-grandsons of Queen Victoria who are named in the pedigree.

free individuals to homozygotes with two copies of the recessive gene is 75:25, which is three-to-one—the same ratio that he found with his smooth and wrinkled seeds. In the particular family shown in Figure 1.2a, two of the children are homozygous for the cystic fibrosis mutant gene and have the disease while one is a carrier and one has two normal copies of the gene. Precise Mendelian ratios only apply to large numbers.

Dominant Gene Inheritance

An example of a family with dominant transmission of a gene is shown in Figure 1.2b. The defective gene in this case results in familial hypercholesterolemia (FH). As the rather unwieldy name suggests, the main clinical problem in this case is abnormally high levels of cholesterol in the blood. So FH itself is not a disease, but rather an inherited metabolic disorder that can lead to disease—in particular, cardiovascular disease. When cholesterol levels remain high, atherosclerosis, or hardening of the arteries, is likely to develop. This accumulation of plaques of fatty material on the inside walls of the arteries increases the risk of coronary heart disease and stroke.

It is the low-density lipoproteins (LDL, the "bad cholesterol") that accumulate in FH and which are particularly damaging to the arteries. Nearly two-thirds of all circulating cholesterol in the blood is carried by the LDL particles. The mutant gene that often causes FH encodes the receptors on the surface of cells that normally take up LDL from the blood. These are called LDL receptors, and they normally function to keep the level of blood LDL at the right level. Even if only one of the two LDL receptor genes is defective (the person is heterozygous), this is enough to reduce the number of LDL receptors by 50 percent, which in turn is sufficient to increase the LDL levels in the blood by three- to five-fold, increasing the risk of cardiovascular disease.

Figure 1.2b shows why such defective genes are known as "dominant," for the simple reason that the FH condition appears wherever there is a half-black symbol—in other words, whenever someone, male or female, possesses a single copy of the defective gene. Unlike cystic fibrosis, in which the amount of chloride channel protein resulting from one normal gene is enough for cells to function properly, in the case of FH the reduction of about 50 percent in the number of LDL receptors does not leave enough for normal function.

X-Linked Gene Inheritance

The examples of inheritance given so far all depend on genes located on what are called the "autosomal" chromosomes, the twenty-two pairs that are not sex chromosomes. But occasionally defective genes are found on the X chromosome, which leads to a quite distinct pattern of inheritance, illustrated in Figure 1.2c, that would certainly have puzzled Mendel.

The royal family of the British Queen Victoria shown in this figure has members suffering from hemophilia A. The defective gene in this case encodes for Factor VIII, one of several protein factors that are essential in the cascade of events that results in rapid blood clotting. Hemophilia was known in earlier centuries in Europe as the "royal disease" because it was so prevalent in the European royal families. Passed on by Queen Victoria, the mutation spread by marriage to the royal families of Spain, Germany, and Russia. Fortunately boys with the disease can be treated using Factor VIII purified from blood or, more efficiently, with Factor VIII made by genetic engineering.

Figure 1.2c shows why females, who have two X chromosomes, are protected against hemophilia when the Factor VIII gene on one of their X chromosomes is defective. As with the recessive pattern of inheritance, the normal Factor VIII gene on the other X chromosome makes all the factor that is needed for normal blood clotting. But the situation with males is like the dominant pattern of inheritance. Since the males have only one X chromosome, there is no backup, and a defective gene on the single X chromosome therefore results in hemophilia in the male. So in X-linked diseases, females are the unaffected carriers, and the males suffer. If the males reach reproductive age, then they will pass on their defective X-linked gene to any females, who then become carriers in turn, but not to males, because males inherit their X chromosome only from their mothers. This explains the Talmudic injunction that sons of the hemophiliac's mother's sisters were exempt from circumcision, but not sons of his mother's brothers. If it's any comfort, the situation is the opposite in birds, in which the females have to do with only one X chromosome, whereas the males have two—but then birds don't have to worry about circumcision. In humans very few diseases show this X-linked pattern of inheritance. In addition to hemophilia, they include various forms of muscular dystrophy, as well as red-green color blindness.

Genes and the Structure of DNA

So far we have, like Morgan, portrayed genes as beads strung out on a string, the piece of string in this case representing one chromosome. This picture, plus the fact that chromosomes come in pairs, has been quite sufficient up to this point to explain the patterns of inheritance for around six thousand different genetic diseases.

Each chromosome is in fact a single molecule of double-helical DNA, packed together with proteins that are critical for its function. In later chapters we consider what all those proteins do, but first we need to see how the genes are incorporated into this double-helical structure, and so we need also a basic grasp of that structure itself. Fortunately the structure is simple, elegant, and quite easy to explain, which is how Watson and Crick were able to present it in just a single page of the journal *Nature*. Here we take a little more space, but not much more, helped by Figure 1.3.

Imagine that you are at the bottom of a spiral staircase. Each step has a height of 3.4 feet, a little more than a yard, so you have to stretch up a bit to make the steps. Each step is rather wide—in fact,

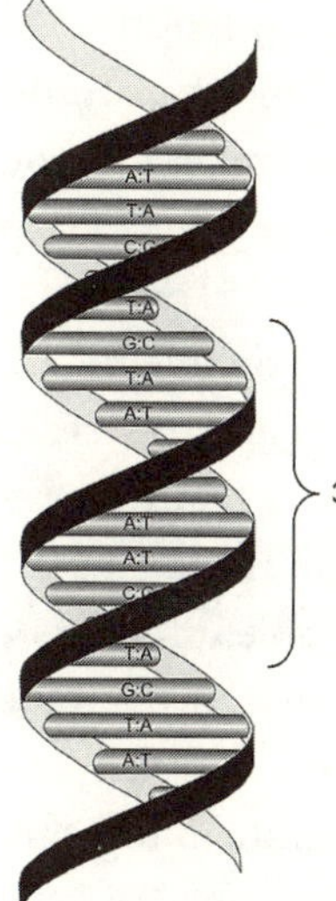

FIGURE 1.3. The DNA double-helix. Four nucleotides—adenine (A), thymine (T), guanine (G) and cytosine (C)—serve as the basis for the genetic code. A binds exclusively with T, and C with G, so allowing faithful replication of DNA.

20 feet wide—and each time you step up, you start to twist around the center by 36 degrees. (We'll see why the steps are this wide and this high in a moment.) As you go up the steps, you notice that each one is labeled "hydrogen bond," so they seem pretty uniform. But then you notice that the steps all the way up are fixed on each side into chemical structures called either purine or pyrimidine nucleotides. The purines are either guanine or adenine (G or A), whereas the pyrimidines are either cytosine or thymine (C or T). And as you continue to count the steps, you notice a pattern. If the step is joined to a G on the right side, then it's always joined to C on the left. Go up a step and it may be joined to an A on the right side, in which case the step joins up with a T on the left. In other words, a purine always pairs with a pyrimidine: they seem made for each other. For short, these are often known as "bases" (A, G, C, and T), which come in "base-pairs" (AT and GC).

Taking out a notepad, jot down the labels on the different nucleotides as you climb upward. Starting from the bottom of the staircase, those from the right read ATGTACAAGGATGTGCT-ATTGTAA and onward, so those on the left read TACATGTTC-CTACACGATAACATT. The sequence of letters seems to make no rhyme nor reason as you jot them down. After exactly ten steps, or 34 feet precisely, the spiral makes a complete turn. In other words, as you look down from the right-hand end of step number ten, you can see someone else standing below on the right-hand end of step number 1.[16]

By now you should be getting the picture of the double-helical structure of DNA illustrated in Figure 1.3. The only difference, to make picturing the metal spiral stairs a bit easier, is that we have made one angstrom equal to one foot. An angstrom is actually quite small, only one one-hundred-millionth of a centimeter, or 0.000000004 inch. So in reality the vertical distance between the purines and pyrimidines that make up the double-helix on each side is just 3.4 angstroms, and the hydrogen bonds join them together (like the steps) across a total diameter of 20 angstroms. To keep that distance exactly equivalent at each step, a purine has to link to a pyrimidine. Purine-purine pairs would be too wide, pyrimidine-pyrimidine pairs too narrow; either way such pairing would lead to a kink in the helix that would make it unstable.

Each chromosome contains one spiral staircase, or one long double-helical DNA molecule. By convention the human chromosomes are numbered from 1 to 22 according to size, with number 1 the largest and 22 the smallest. Chromosome number 1 contains about 248 million base-pairs or 248 million steps in our staircase analogy.[17] That's a very long way up if you're climbing up a stairway. Based on 3.4 angstroms per step, this comes to a bit more than 3 inches long, or more than a yard long if you add up all the chromosomes together, totaling 3.2 billion base-pairs, the same number of letters as two thousand copies of *War and Peace*. You might wonder how such a huge collection of chromosomes could fit into the tiny nucleus of a cell, and the answer is a marvel of packing. Instead of being stretched out lengthwise in the way pictured here, in reality the chromosomes are tightly folded, making your vacation packing of too many items in one crammed vehicle look trivial by comparison.

In the DNA double-helix, the hydrogen bonding is of exactly the right strength for the molecule. It's like a zipper on a jacket—fixed enough to hold the garment together, but not so fixed that unzipping becomes too difficult. As Watson and Crick pointed out in their 1953 paper in *Nature*, DNA has exactly the right structure to facilitate its own replication. It simply unzips down the middle, and then each strand of the helix is used to generate a complementary strand in which the information is replicated by keeping to the exact same GC and AT pairing rules. The outcome is two identical daughter molecules of DNA.

The nucleotide sequence recorded here does indeed encode the real sequence of a part of a protein. Twenty different amino acids are used to build different proteins, and the specific sequence of these amino acids gives the protein its particular properties.[18] Each amino acid is encoded by a triplet of nucleotide base-pairs in the DNA sequence known as a "codon." Using our code manual,[19] we can take the sequence and now separate it out into triplets:

> ATG-TAC-AAG-GAT-GTG-CTA-TTG-TAA

and immediately translate that sequence into:

> START-Tyrosine-Lysine-Aspartate-Valine-Leucine-Leucine-STOP

where each of the names shown refers to a particular amino acid, and the START and STOP signals show the code-copying machinery where to start and stop reading. In fact, the START signal ATG also encodes methionine, so this amino acid is placed at the start of a protein, but is often clipped off later. The fact that both CTA and TTG encode for the amino acid leucine illustrates the fact that the code contains redundancy: in some cases, several triplet codons encode the same amino acid. Only six amino acids are shown in this sequence, far fewer than would be found in a typical protein, which might typically contain hundreds of amino acids or more. A section that is only a few amino acids in length is called a "peptide." But the example illustrates an important principle: how the sequence of DNA is converted into the sequence of amino acids in a protein. Since proteins give the structure and function to living organisms, we can now see how DNA can operate as a storehouse of information for building bodies, and we will find out in more detail in the next chapter how that happens.

Note that as we ascend the spiral staircase, only the right-hand strand of the DNA staircase can be used to read off the genetic code used to make proteins. The right-hand side is called the "sense" strand of DNA. The left-hand strand of DNA is definitely not "nonsense" by contrast, because it contains the precise complementary sequence of nucleotides. The left-hand side is the "antisense" strand, and it plays a critical role in conveying the genetic information out of the nucleus.

We can now see why biologists greeted with such delight the publication of the DNA structure. Mendel's discovery of particulate inheritance, followed by Morgan's idea of genes as beads on a string, could now be given a specified molecular definition. For several decades from the 1950s onward, the gene was seen as a specific sequence of DNA that encodes a protein, or a section of a protein (called a "polypeptide"). The START and STOP signals made it clear where the gene was to be found. The complete nucleotide sequence contained in the DNA of any organism is known as its "genome," and the complete sequencing of human DNA has revealed that our own DNA contains about twenty-one thousand of these protein-encoding genes. Yet as more and more secrets of DNA are brought to light, the clearer it becomes that genes are not simply beads on a string. Life is more complicated, and only when we understand a little more of that complexity can we appreciate the very varied ways in which information flows from the genes to build bodies.

CHAPTER 2
Genes and Information Flow

FRANCIS CRICK introduced the idea of a "central dogma" into genetics in 1958.[1] As he later confessed, Crick was not sensitive to the accepted meaning of the word "dogma," and its use caused "almost more trouble than it was worth." But although the word "dogma" is not often used in science, the core of the idea has withstood the test of time. Basically it says that information flows from genes to proteins, but not the other way around. The flow works as shown in Figure 2.1.

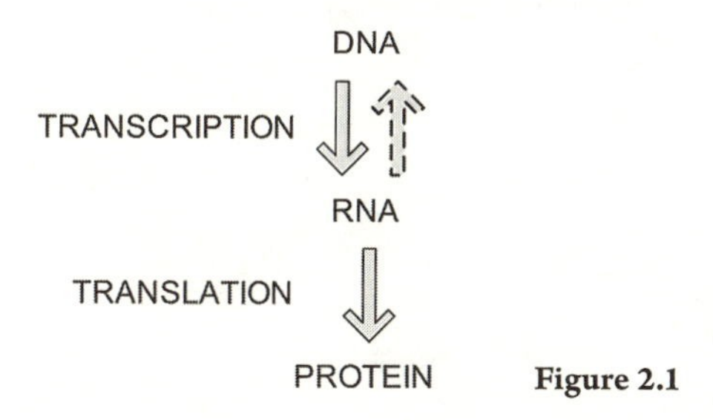

Figure 2.1

RNA is a chemical very similar to DNA in structure and contains the same nucleotides as DNA, except that the pyrimidine uracil (U) is found in place of thymine. RNA can therefore contain information just like DNA by means of the same kind of a specific sequence of nucleotides. RNA can also form a double-helix like DNA, but in this present context only a single strand of RNA is involved.

What happens is that the specific sequence of nucleotides on one of the two strands of double-helical DNA in the nucleus of the cell is "transcribed" into a molecule of single-stranded RNA, known as messenger RNA, or mRNA for short. The message encoded in the RNA is then "translated" into the amino acids of a particular protein. Note how the terminology of language is used so often in talking about the way that information flows from the genes. "Transcription" sounds as if there were a very small DNA clerk sitting inside the nucleus and calling out the sequence for the RNA clerk to copy down. The reality is a bit more sophisticated than that.

Think back to the DNA helical staircase. The right-hand strand going up was the sense strand, in which the genetic code made sense as it was being read from the bottom upward. The challenge now for the DNA is how best to convey its information out of the nucleus to the cell cytoplasm, where the information is needed to synthesize new proteins. The DNA accomplishes this task in an elegant and essentially simple way using an enzyme called an RNA polymerase, which "reads" the antisense strand. In front of the gene is a "promoter region" that tells the RNA polymerase where to begin:

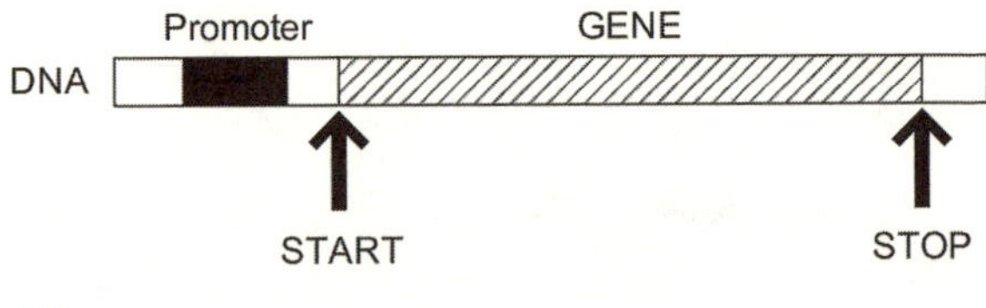

Figure 2.2

Once certain regulating factors, known as "transcription factors," do or do not bind to the promoter region, the gene is then ready to be transcribed; the polymerase begins to copy the DNA into mRNA somewhat before the translation START site, which is identified by the triplet nucleotide sequence ATG. As the polymerase shuttles along the DNA, it copies each DNA nucleotide one by one into the complementary nucleotide sequence of the mRNA, using the same pair-wise rules already outlined: a G in the DNA produces a C in the mRNA, and an A produces a U (not a T because RNA uses U in place of T). Once the polymerase gets to a STOP triplet codon, it falls off and trundles off to work on another gene. Using the same example sequence as in chapter 1, we can see in Figure 2.3 what our mRNA looks like.

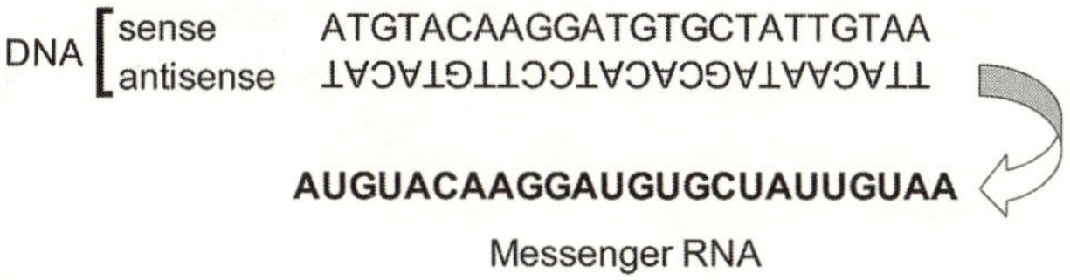

DNA [sense ATGTACAAGGATGTGCTATTGTAA
antisense TACATGTTCCTACACGATAACATT

AUGUACAAGGAUGUGCUAUUGUAA

Messenger RNA

FIGURE 2.3

Inspection of the mRNA shows that it now has exactly the same sequence as the sense strand of the DNA, except that it has a U in place of each T in the DNA. Once an mRNA transcript is made in this way, the gene is said to be "expressed." The transcript is now ready for work—to use its information to encode the amino acid sequence of a protein in a process called "translation," described below.

First, though, notice that the example above is not really a gene, but just a small portion of a gene used for illustration. In reality, as already mentioned, genes encode proteins that are several hundred—or sometimes more than a thousand—amino acids long. A human muscle protein called Titin is an astounding 34,350 amino acids long,[2] so it takes 3 x 34,350 = 103,050 nucleotides to encode it. At any given time, hundreds of different mRNA molecules of all different sizes are being generated in the nucleus. The transcription machinery is incredibly fast; on average, RNA polymerase adds 40 new nucleotides to a given mRNA every second.[3] This means that a gene with 1,000 nucleotides is transcribed into a new molecule of mRNA in only twenty-five seconds, which is pretty impressive.

We must also not lose sight of the dotted arrow that goes from RNA to DNA in our central dogma (Figure 2.1). An enzyme that can catalyze the formation of DNA from RNA was discovered in 1970. Called reverse transcriptase, it plays a key role in the way that certain viruses infect cells. Viruses consist of little packages of either DNA or RNA surrounded by a few proteins that vary in number depending on the type of virus. They can exist alone, but need to infect cells to multiply like a parasite, taking over the machinery of the cell to make more copies of themselves. Some RNA viruses, called retroviruses, carry a tool kit of proteins, one of which is reverse transcriptase, which they use to convert their genetic information into the cell's own DNA. A well-known example of an RNA retrovirus is HIV, which causes AIDS.

Except during some viral infections and in a few other circumstances, the flow of information going from DNA to RNA is what really matters, and we now need to see how the mRNA is translated into the specific amino acid sequence of a protein, one of the most amazing processes in the whole of biology.

Not Lost in Translation

Once the RNA polymerase falls off, the completed mRNA moves out of the nucleus through little holes into the cytoplasm of the cell (see Figure 2.4). There it is used as a template to construct the new protein with the help of sophisticated pieces of molecular machinery called ribosomes.

Clearly for a translation process to work properly, a translator is required, and in this case a collection of clever adaptors called transfer RNAs, or tRNAs, carry out the work. The tRNAs contain the same four nucleotide bases as the other types of RNA, such as mRNA. But they are different from mRNAs in being much shorter and of a standard length, usually 74 to 95 nucleotides. Furthermore, they have a standard, rather beautiful cloverleaf-like struc-

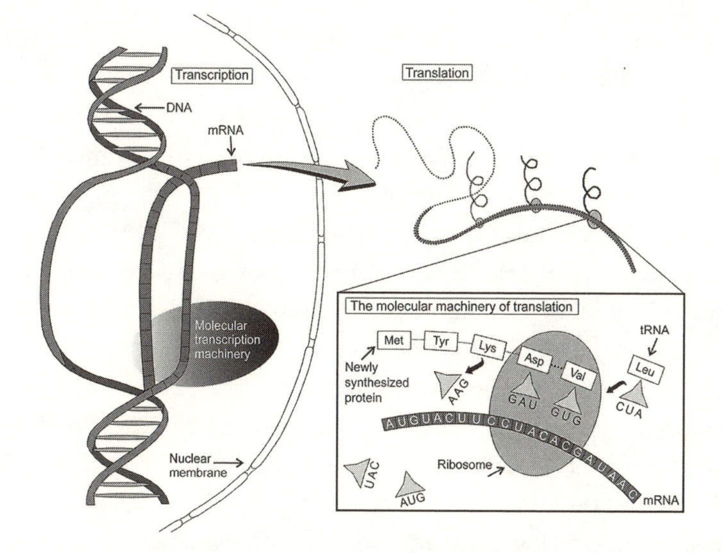

FIGURE 2.4. Transcription and translation. Transcription is the process by which RNA is synthesized using a DNA template. Transcription occurs in the nucleus of the cell, and mRNA molecules then move into the cytoplasm, which occupies the main body of the cell, where they are translated into proteins. Translation is the process by which proteins are synthesized from individual amino acids using mRNA as a template. Translation occurs on ribosomes, which bring together the mRNA and amino acids bound to tRNAs.

ture. Once again the specific nucleotide pairing comes in handy to arrange the various leaves of the clover.

Each tRNA includes an acceptor "arm" that attaches by a chemical bond to a specific amino acid, as Figure 2.4 illustrates. At the other end of the tRNA molecule is another arm, and right in the middle of it is a triplet sequence of nucleotides known as the anticodon. The anticodon corresponds exactly to the codon in the mRNA sequence, like a plug fits into a complementary socket. Using an example from Figure 2.3, the AUG codon in the mRNA (equivalent to the ATG codon in the sense DNA strand) binds to the UAC anticodon in the tRNA, and sure enough we find methionine bound to the other end of this particular tRNA. The tRNA that binds methionine serves as the START signal for the protein.

The genetic code specifies twenty amino acids, but more than twenty different tRNA "plugs" exist because there are more than twenty different mRNA "sockets." This relationship of several different codons used to encode a single amino acid is known as the "redundancy" of the genetic code.

The clever ribosome now plugs in the next tRNA, which has AUG as its anticodon, corresponding to codon UAC in the mRNA, and this time we find the amino acid tyrosine bound to the acceptor arm of the tRNA. The first two amino acids of our new protein are now next to each other, brought into position like two boats moored side by side. An enzyme then catalyzes a peptide bond between the two amino acids, which holds them together quite firmly, and the ribosome shuttles along to pick up the next amino acid in the growing chain until we have the sequence: methionine-tyrosine-lysine-aspartate-valine-leucine-leucine, as introduced above. As mentioned, a real protein is much longer than this. On the machinery goes until it finally reaches a STOP signal, and the completed protein drops off.

By this time the protein will already have folded up into a shape that its precise amino acid sequence specifies, and this specified conformation gives the protein its particular properties; it may be an enzyme, a structural protein, a regulatory protein, a specialized muscle protein, or whatever else the body might need at that particular moment.

The same basic process of transcription followed by translation, and virtually the same genetic code, are common to all living things: bacteria, plants, and animals. Bacteria are different from plants and animals in that they don't have a nucleus, so the mRNA need not travel from the nucleus into the cell. They often have just one main chromosome, compared, for example, to our twenty-three pairs of chromosomes, which helps simplify the process in bacteria. Other minor differences occur in the way the system works in cells with or without a nucleus, but the general idea is the same.

Doing some number crunching in bacteria helps us to appreciate just what a remarkable set of processes the flow of information from DNA to mRNA to proteins involves. Given the soup of hundreds of chemicals in the same cell, how on earth do they find each other? The numbers give us some clues. Take *E. coli* as a typical bacterium; no less than 30 percent of its dry weight is committed to the molecular machinery concerned with the flow of information from genes to proteins.[4]

At any one time, there are about fifteen hundred molecules of mRNA in the cell, encoding the message from about six hundred different genes, so there may be only two or three copies of any one mRNA specific for a particular protein that needs to be synthesized. But there are also twenty thousand ribosomes and about three thousand copies of each tRNA, plus a large excess of amino acids swilling around to keep the protein synthesis factory going. So cells maintain a large excess of the raw materials needed to construct something new, while the instructions are kept relatively few. In the case of *E. coli*, the information flows from one gene in one molecule of DNA (its single chromosome) to a few mRNA molecules, which, in the presence of two hundred thousand tRNA molecules and millions of amino acid molecules, are then used as templates to construct one particular protein. The system is incredibly sophisticated.

The Changing Notion of the Gene

The advent of DNA and molecular biology meant that a gene was no longer merely a conceptual idea invoked to explain the way in which traits were inherited, but rather an actual stretch of DNA of a defined length. For several decades after the central dogma was first introduced in the 1950s, the idea of a gene had a comfortingly clear definition: an inheritable stretch of DNA that encoded a single polypeptide or protein, transcribed and translated as we have just described. This is how we can say that human DNA contains about twenty-one thousand protein-encoding genes.

But as more and more genomes have been sequenced, a rather puzzling fact has emerged. The number of genes encoding proteins, defined in this traditional sense, appears to bear little relationship to the complexity of the organism. True, bacteria that have only a single cell tend to have genomes containing five thousand genes or less. But the number of human genes is similar to that of the sea urchin (about twenty-three thousand) or the nematode worm (about nineteen thousand), and is a lot less than the humble ciliated protozoan *Tetrahymena*, familiar to all those who like peering down microscopes to examine pond water, which has about twenty-seven thousand genes. To top them all, rice has about fifty-seven thousand![5] Furthermore, the twenty-one thousand protein-coding genes in human DNA represent only about 1.5 percent of the 3.2 billion nucleotides found in human DNA taken as a whole. What is going on?

Life is never simple, and what we now know is that the information flow from DNA to the rest of the cell takes many different forms. Much recent investigation has shown that RNA molecules may act as end-product messengers in their own right, not just as intermediates in the production of proteins. Furthermore, a single gene can make several different proteins, in contrast to the "one gene–one protein" idea of a previous era. All this means that we no longer have any universally accepted clear definition of a gene. As soon as a clever biologist comes up with a definition that looks good, another biologist finds another way in which DNA encodes information, rendering the neat definition obsolete.[6]

For our present purposes we can simply define the current understanding of a gene as "any inheritable DNA sequence that encodes information." The three examples that follow show how this works in practice in terms of information flow, and the next chapter describes how genetic information is used to build and regulate living organisms. The first example shows how one gene can generate multiple units of information; the second example describes how small RNA molecules are like the cell's e-mail system; the third example probes into the mysterious dark matter of the genome. These examples don't cover all the various ways in which information flows from genes to cells, but they cover quite a lot.

One Gene, Many Proteins

One of the first findings to shake the "one gene–one protein" idea was the discovery that one protein-encoding gene can be used to make several proteins through a process called "alternative splicing." To understand how this works, we need to go back to the account of transcription given in the previous chapter. For the sake of simplicity, one of the steps in the process was skipped in the previous figure. The more complete picture is shown in Figure 2.5.

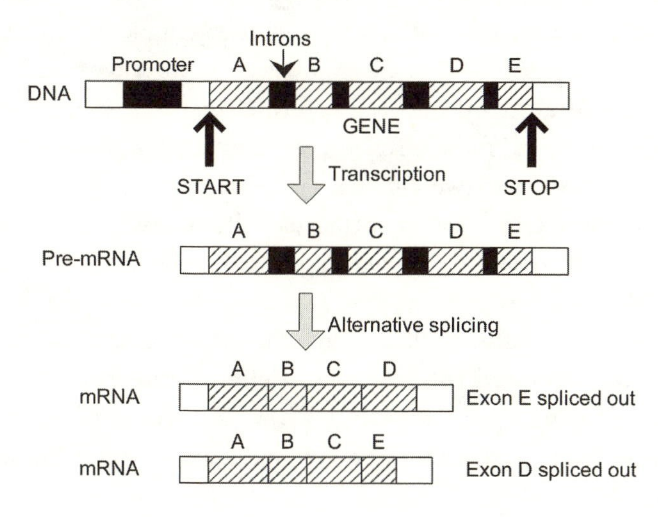

Figure 2.5

Protein-encoding genes are divided into stretches of DNA called introns and exons. Figure 2.5 shows five exons in the gene, labeled A to E. Four black introns appear between the exons. The exons contain the coded nucleotide sequence that will eventually be translated into the amino acid sequence in the protein. The complete stretch of the gene, including both exons and introns, is transcribed to generate the immature form of the mRNA, known as pre-mRNA. But while the pre-mRNA is still in the nucleus, it undergoes a processing step whereby the introns are spliced out and the exon sequences are joined up together to form the mRNA, which then goes off to the cytoplasm to be used as a template for translation, as already described. Once the exons are joined up, they form a correct and continuous sequence of triplet codons; you can't tell where the introns have been spliced out. We can say that the protein-encoding sequence of the mature mRNA message can be read as a complete sentence in one breath, without any commas or semicolons.

The beauty of this system is that it allows for different combinations of exons to generate different mRNAs, and therefore different proteins with different functions. In the example shown, exons A to C are invariant—they are always used—whereas exons D and E are variable. Two mRNAs can be formed that contain A to C plus either D or E, and these mRNAs then shuttle off to the cytoplasm to be translated into two different proteins.

A nice example of the importance of alternative splicing comes from the little fruit fly *Drosophila,* generally only 2 to 4 mm long, which has the best-characterized genes of any complex multicellular organism. Male fruit flies often have sex in mind, breeding fast and often. Provided with a healthy diet of rotting fruit, they can produce the next generation within a week or so, which is one of the reasons they are such a popular model organism for the study of genetics.

Males also have the largest sperm cells in the world. The most common species of fruit fly used for genetic studies, *Drosophila melanogaster,* has sperm cells about 1⁄14 of an inch (1.8 mm) long, which may not sound very big, but makes them three hundred times longer than human sperm. One species (*Drosophila bifurca*) has made it into the *Guinness Book of Records* for its sperm, which measures over 2 inches (5.8 cm) long, about twenty times longer than the fly that produces it. Most of this length is taken up by a very long sperm tail, which is relatively easy to pack into a tight space.[7]

With this kind of impressive kit on board, it's not surprising that the males are keen to mate, but nothing gets off the ground without courtship. Upon spotting a female, the male performs a carefully choreographed dance, tapping on the female's legs to gain attention and singing a courtship song by vibrating one wing. The interactions are helped along by both partners' pheromones, chemicals detected by special receptors that signal that sex is in the air (although you might think that leg tapping alone would do the trick). The female courtship behavior is mainly characterized by running away, but if all goes well, copulation finally results.

In some males, though, all is not well. They have a gene mutation called "fruitless" (or *fru*; geneticists love picking names for their mutants, and they are usually then given three-letter abbreviations).[8] A mutation simply means a change in the gene sequence, which in this case results in an abnormal protein. The *fru*-mutant males engage in early courtship behavior with both males and females indiscriminately and leave out altogether the later stages of courtship like singing or copulation. Not surprisingly, the *fru*-mutant males are sterile. Clearly the normal *fru* gene is a key regulator of a set of male behaviors involved in courtship, and as it's only active in about five hundred of the one hundred thousand neurons that constitute the *Drosophila* "brain" (more formally, its central nervous system), those five hundred neurons must be important in regulating courtship.

The *fru* gene is also active in the normal female *Drosophila* brain, even though female flies engage in a very different kind of courtship behavior than males do. While *fru* is the same in both sexes, the Fru protein in males is slightly different than in females, which is explained by alternative splicing. The *fru* gene encodes a transcription factor, one of those key regulatory molecules that binds to the promoter regions of other genes to switch them on or off. The *fru* gene pre-mRNA is alternatively spliced. In the male this generates a mature mRNA that in turn generates a transcription factor that binds to the promoters of many different genes in the five hundred or so neurons already mentioned, so acting as a master switch that triggers a whole sequence of events leading to male courtship behavior. By contrast, in the female *Drosophila,* alternate splicing generates an mRNA that in turn generates a protein which does not have these effects. Mutations in the *fru* gene in the male, but not in the female, can therefore block courtship behavior. Without understanding alternative splicing, putting together all the pieces of this fascinating molecular and behavioral jigsaw would be difficult.

Some genes have exceptionally large repertoires of exons that can be spliced in alternate ways. Once more the fruit fly gives us a prize-winning example. A *Drosophila* gene named *Dscam,* involved in guiding nerve cells to make the right connections, can potentially generate up to 38,016 different variant proteins by alternative splicing of its different exons.[9] This number is larger than the total number of protein-encoding genes present in *Drosophila,* and provides a rather dramatic example of the way in which a relatively small number of genes can be used to generate a huge range of protein products.

In the human, as much as 95 percent of all our genes that contain multiple exons are alternatively spliced just like the fruit fly genes. DNA contains a code within a code in the sense that certain features identify the introns and exons where splicing occurs. But this code is a lot more complicated than the sixty-four triplet nucleotides that encode the amino acids. Instead it consists of a complex algorithm that includes more than two hundred different features of DNA structure that predict where splicing will occur.[10] Although our own twenty-one thousand genes may not seem very many to provide the instructions to build something so complex as a human body, when we realize that up to 95 percent of these genes can generate different variant proteins, in many cases having different functions, then the number of proteins that our genome generates begins to look more respectable—certainly more than one hundred thousand. In addition, once proteins have been synthesized, they can be further modified by more than ten different kinds of post-translational modification, whereby other chemical groups such as phosphates are added to specific amino acids within their structures. So more than a million different functional proteins can come from our modest repertoire of around twenty-one thousand genes. The situation seems very similar in other living organisms.

Small RNA as an Information Carrier

Our second example of information flow from genes to cells takes us into the world of RNA. One of the biggest revolutions in our changing understanding of the gene has come from the discovery that many different types of small RNA molecules can communicate information from DNA either within the cell, or indeed all the way round organisms such as plants from cell to cell, as has been shown in plants.[11] Unlike the situation with mRNAs and tRNAs, these types of RNA are not simply mediators or adaptors to convert DNA information into protein information, but rather are direct communicators of genetic information. If the mRNAs and tRNAs together are equivalent to the servers that transmit messages around the Internet, then the small RNA molecules are more like the actual e-mails that wing their way directly from our minds and fingers to a recipient mind. Our intention here is not to give a comprehensive survey of all the different types of small RNA, but instead provide some examples of the fascinating ways in which they convey their messages.

Why are our brains so much bigger than that of our nearest relatives, the great apes? Since we last shared a common ancestor with the chimpanzee about 6 million to 7 million years ago, the human brain has expanded in size many-fold, one of the most rapid and striking changes to occur in evolutionary history. Most striking is the tripling of brain size that occurred about 1.5 million to 2 million years ago in the human lineage—not overnight, but rapidly enough that it attracts our attention. Today the brain size of the great apes is in the region of 300 to 500 grams, even though their body size is comparable with ours, whereas the average human brain weighs in at 1,300 grams. To find out why, those working in the field of bioinformatics—computer nerds who specialize in tackling genetics—compare in minute detail our genomes with those of our nearest cousins. They are like detectives, looking for the genetic clues to uncover the key differences between the brain of a human and the brain of a chimp.

One way of approaching such a challenge is to identify common genes expressed in the brains of humans, great apes, and other species, and then clock the speeds at which these genes have changed over the 6 million or so years that separate us in evolutionary time. Among regions of DNA that are very similar ("highly conserved") among all mammals and thus easily compared, forty-nine have been identified as human accelerated regions (HAR) because they have changed so quickly in the human lineage. Winner of the race in terms of the amount of change it incorporates, meaning the number of changes in its nucleotide sequence, is a stretch of DNA called HAR-1. Much to the surprise of the investigators, HAR-1 was found not to encode a protein, but instead to represent one of the new "RNA genes."[12] Even more surprising was their finding that all but two of the HAR regions identified lie outside the "traditional" protein-encoding genes of the DNA.

Being an RNA gene means that the DNA in the nucleus of the cell is transcribed just as described above, but the RNA transcript is processed differently than pre-mRNA and generates one or more shorter stretches of RNA. In turn these shuttle out of the nucleus into the cytoplasm to exert their various regulatory effects without any need to be translated first into proteins.

The RNA gene HAR-1 is just 118 nucleotides long, and yet there are eighteen changes in the nucleotides between us and the chimps, whereas pure chance would have predicted less than one change.

The data suggest that these changes took place in the human lineage, not in that of the great apes, as only two nucleotide differences are present between the HAR-1 sequence of the chimp and the chicken, which last shared a common ancestor at least 310 million years ago. No HAR-1 gene at all has been found in frogs, fish, or invertebrates, suggesting that the gene originated not more than 400 million years ago.

The precise function of HAR-1 is not yet known, but its pattern of expression in the developing human brain—in particular the neocortex, which is involved in higher cognitive functions—suggests that it is suitably poised to influence the migration of neurons (brain cells) as the brain develops. HAR RNA genes tend to be located in the DNA near protein-encoding genes that are known to be important in the development of the nervous system. Based on the actions of other RNA communicators, the HAR family likely consists of regulatory genes that switch protein-encoding genes on or off.

These newly identified RNA communicators are generally classified as either long or short "noncoding" (nc) RNA and may be written as "ncRNA" to distinguish them from mRNA and tRNA, with which we are already familiar. The word "short" here is a bit arbitrary, but generally means less than 200 nucleotides long. The term "noncoding" simply refers to the fact that unlike mRNAs, these RNA molecules do not encode proteins. The "nc" tag is a bit of a put-down because in reality these molecules do a great job encoding precious information; they just do it differently than the other guys down the DNA road. Altogether 8,483 RNA genes have been reported in the human genome, compared to around 21,000 protein-coding genes.[13]

One large and important class of RNA communicators is microRNAs, which are only 18 to 24 nucleotides long. They are found in animal and plant cells, and even in algae, making them very old from an evolutionary perspective, although they have not yet been described in bacteria.

As explained above, microRNAs are made from longer RNA molecules that are trimmed down after transcription to generate the final 18 to 24 nucleotide products. About one thousand of these RNA "e-mails" have been identified in the human genome, although precise functions have so far been assigned to only some of them. Many of them are found within the gene introns described already. If you were still wondering why protein-encoding genes bother to have introns, given that they are spliced out to generate the mRNA, then here is one answer: some introns are used to encode these key regulatory RNA molecules. It's a case of two for the price of one: the gene gets you a protein as well as a regulator thrown in for free.

MicroRNAs are able to operate as they do because they consist of complementary nucleotides to those found in specific stretches of mRNA, meaning that as usual a G pairs up with a C, and an A with a U. Once processed and out in the cytoplasm, they can target specific mRNA molecules and prevent their translation by two different mechanisms—either inhibiting the way that the mRNA is translated, or targeting the mRNA itself and recruiting proteins that cause its degradation. This second mechanism is the most important, but either way the amount of protein product from the protein-coding gene is reduced. So microRNAs act as repressors of gene expression.

Estimates hold that each microRNA might regulate tens or even hundreds of protein-coding genes, so probably more than one-third of the "traditional genes" in the human genome are regulated by these upstarts. It's as if the cell had a whole range of subversive PIN numbers at its disposal that it could send out to prevent transactions just at the moment when the money is about to be drawn from the mRNA ATMs. MicroRNAs therefore provide the cell with an incredibly fine-tuned regulation system.

This kind of regulation by microRNA is going on millions of times in our bodies every second of the day, keeping hundreds of different systems finely tuned. In fact, we now know that micro-

RNAs are involved in regulating aging, cancer, metabolism, and muscle size. Every known cellular process is regulated to some extent by microRNAs. Like e-mails, you cannot get away from them.

Long Noncoding RNAs

One of the greatest surprises in the recent investigation of our own DNA and that of other species is that up to 90 percent of all our DNA is transcribed, often from either DNA strand. You may remember that when we described the DNA transcription leading to the mRNAs that encode proteins, we were careful to point out that transcription takes place from the antisense strand of the DNA. If transcription were from the sense strand, then the mRNA would no longer be able to act as a template for protein translation, because translation would be attempted back-to-front and nonsense would result. It would be like spelling "thecatsatonthemat" as if it were "tamehtnotastaceht," which wouldn't make any sense at all. But long noncoding RNA can be transcribed from either DNA strand.

In cosmology it is well known that 95 percent of the universe consists of either dark matter (25 percent) or dark energy (70 percent), invisible but predicted to be there to satisfy certain mathematical and observational predictions. In like manner, if we take the 1.5 percent of protein-coding genes and approximately 3 percent of associated regulatory genes as the known portion of the human genome, then the term "dark matter" could apply to the 95 percent of the genome that has an unknown function, or maybe no function at all. Indeed, in earlier times the portion of the genome that did not encode protein-coding genes was rather unwisely labeled "junk DNA," but that phrase is rarely used today with the realization that much of the genome is transcribed, yielding a complex network of tens of thousands of overlapping RNA transcripts representing much of the 3.2 billion nucleotides in our genome.

As soon as these results became apparent, some scientists ascribed the results to noise. If you have an overenthusiastic transcription machinery in your nucleus, all ready to transcribe the critical protein-coding regions, then it's not surprising if some irrelevant portions of the genome are transcribed as well in the process. It's as if you had a very diligent typist whose task it was to transcribe a one-hour formal discussion from a tape, but who then enthusiastically went on to transcribe also a few hours of cocktail conversation that was incidentally recorded at the end of the tape after the formal part of the meeting was over.

Certainly there is as yet insufficient evidence that all these long ncRNA transcripts mean anything for the function of the cell. Also, newer approaches are suggesting that the actual proportion of the genome that is transcribed to some extent is much less than 90 percent.[14] Despite this, plenty of evidence remains that at least some of the "dark matter" noncoding RNAs are functional. For example, transcription of noncoding RNAs can be regulated during development; they display a pattern of expression specific to certain types of cell, localize in specific places within the cell, and are associated with certain diseases.[15] Furthermore, some are highly conserved during evolution and show evidence of selection pressures.

It turns out that the transcription itself of certain long noncoding RNAs can have a profound effect on promoting the transcription of nearby protein-coding genes. Imagine one student dorm having a very rowdy party (the "long noncoding RNA house"), and then the rowdy party behavior spreads rapidly across the campus to influence the neighboring dorms. Conversely, the opposite effect can occur with other long noncoding RNAs: once transcribed, they cause the "silencing" of nearby genes. This time the "silent long noncoding RNA dorm" spreads its stilling influence to the neighboring rowdy dorms (unlikely in real life, but you never know).

The long noncoding RNAs are also involved in binding to the transcription factors that bind to promoters and trigger the transcription of protein-coding genes, rendering their actions more potent. The parallel action in American football would be the linemen (the two guards, the two tackles, and the center) opening up a hole in the defense so that the running back can run through.[16]

Some long noncoding RNAs also act as precursors for the small ncRNAs already discussed, so they are only long because they haven't yet been chopped up into smaller pieces, each piece sometimes having quite distinct functions. Other long noncoding RNAs interfere with the transcription of other DNA regions. What is quite clear is that at least some of the long noncoding RNAs play important roles in cell function. It will be fascinating to find out more in the future about these tantalizing molecules. But taking protein-coding genes, alternative splicing, and RNA genes together, we can already see that the flow of genetic information from the DNA to the rest of the cell occurs by a highly regulated and complex set of processes.

Imagine that one of your cells, perhaps ten microns in diameter, has been blown up a billion-fold to make it look like a huge factory one hundred meters across. You are sitting in a chair by the nuclear membrane, the porous wall that separates the nucleus from the cytoplasm. Sci-fi images of cities on other planets with people carriers whizzing through the air all around you would have nothing on what you now see before your startled eyes. A hive of frenetic activity is apparent in both directions. On one side, toward the nucleus, the chromosomes are like great writhing masses, with thousands of protein and RNA molecules dangling from them in all directions, and thousands of other molecules whizzing in and out like bees buzzing around a pot of strawberry jam on a warm summer's day. On your other side, toward the cytoplasm, the scene is no less frantic, with millions of different molecular events going on in one great bewildering yet highly coordinated network, with a bombardment of new instructions constantly arriving from the big bosses in the nucleus.

What we now need to do is see how this great genetic information flow coordinates together to build and regulate bodies. And is it really just the DNA that is the big boss?

CHAPTER 3
Body Building and Genetics

THE FIFTEEN-MINUTE RACE has been an uphill battle—only a few thousand sperm out of hundreds of millions have made it to the final stretch. Swimming furiously up the fallopian tubes at a few millimeters per second, helped along by contractions of the tubes' walls, millions die along the way in one of the world's most competitive marathons. The remaining sperm plunge their little heads into the outer layer of the egg wall, releasing enzymes that weaken the tough ramparts. Finally a single sperm manages to penetrate the wall and, within seconds, reaches the inner membrane layer that surrounds the egg's cytosol. There it fuses its complete contents with the egg, so that sperm and egg become a single cell. Within a few more seconds, other enzymes are released from the cytosol to render the egg wall completely resistant to any further interlopers, and the wall remains intact for another five days yet, just to make sure. A few other sperm that make it to the finishing line knock their heads on the egg wall in vain.

Each of our lives began this way. In addition to their complement of twenty-two nonsex chromosomes, about 50 percent of sperm contain a single X chromosome, and the other 50 percent a single Y chromosome. The egg always contains a single X chromosome, so the sperm that fertilizes the egg determines the sex of the embryo (remember that females are XX and males XY). Had another sperm swum just that tiny bit harder, then you could so easily have been male rather than female, or vice versa, but then of course "you" would have been someone else altogether.

Once fertilization occurs, the single chromosomes from sperm and egg pair up to form the pairs of chromosomes that are now going to be present in every cell of the body that results. Apart from sperm having either an X or a Y, and the eggs having one of the female's two X chromosomes randomly distributed into them, the sex cells contain single copies of all the other twenty-two chromosomes. The sex cells, the sperm and the egg, are called "haploid" because they contain only single copies of each chromosome, whereas the fertilized egg is "diploid" because it contains paired chromosomes. All the "somatic cells," the cells used to build the body, are now likewise diploid from the time of fertilization onward.

The time of fertilization is when we find out who the real boss is. The simple reason is that naked DNA can do nothing without interpretation. By itself DNA would be as useless as a piece of software on a CD without any computer to run it on. In reality DNA is intertwined with proteins, and the DNA-protein mix, called chromatin, makes up the chromosomes. A steady stream of protein messengers, many of them transcription factors, also regulate gene expression.

The key role of proteins in regulating genes is seen most clearly in the early moments following fertilization. Within days, thousands of genes have been regulated by proteins and the fertilized egg has undergone several cell divisions, well on its way to becoming the zygote that will implant into the wall of the uterus six to seven days later.

Proteins also play a vital role in determining which genes will be expressed in the different tissues of the body. The human body contains more than two hundred different cell types, collectively known as "tissues." The tissues are specialized to carry out the different functions of the body, whether brain, liver, muscles, or whatever. Each of the approximately 10^{13} cells of the human body (that's a one with thirteen zeros after it), with the exception of red blood cells, contains a nucleus, which in turn contains the usual diploid twenty-three pairs of chromosomes (except for the eggs or sperm, which contain the haploid twenty-three single chromosomes, as already highlighted). This means that virtually every cell in the body contains the full genetic information to build a complete new body. But each cell type requires only a certain repertoire of genes to be switched on—to be "expressed"—to carry out its specialized set of activities. Because brain cells are complex and need to do lots of different jobs, they need about 60 percent of their total number of protein-coding genes to function properly, whereas skin cells need only about 40 percent to be switched on. Switching on all the genes in every cell would be wasteful, a bit like having all the lights on in a large building when only a few rooms are actually being used. In addition, the cell would no longer be specialized for its particular tasks if all its genes were being transcribed all the time.

The lack of silencing of gene expression is particularly dramatic during the early few days of embryonic development when the first cell divisions are taking place. A few days after fertilization, the embryo has developed into a blastocyst, consisting of a flattened series of cells known as the trophoblast. This series of cells will soon contribute to the formation of the placenta, surrounding an inner cell mass that will develop, all being well, into the new individual. The cells from the inner cell mass are "pluripotent" during this stage of development and are known as "stem cells." They have the potential to develop into any type of cell the body needs, which is why twinning can occur at this very early stage of development. The embryo simply splits in half. Each half develops into an identical twin, containing the same set of genetic instructions, so the resulting twins are of course of the same sex. All the information needed to build a new identical body is found in each half. By day six the blastocyst begins the process of implanting in the wall of the mother's uterus.

Whether the chromatin containing the nuclear DNA is in a very tightly bound or more loosely bound open state determines whether the transcription factors can gain access to switch particular genes on or off. Stem cells have very open accessible chromatin, which is what helps give them their unique properties.[1] As the cells begin to develop ("differentiate") in order to carry out their specialized functions, the chromatin becomes more rigid. Great swathes of the genome are silenced, and other genes are switched on to carry out their tasks in mediating the development of cellular specialization.

These developmental changes in the cells are coordinated in large part by "master genes" that regulate dozens or even hundreds of downstream "slave genes," which switch on or off in obedience to their commands. For example, a master regulatory gene called Chd1,[2] which encodes a chromatin remodeling enzyme, is responsible for maintaining the chromatin in its open form, and consequently maintaining stem cells in their pluripotent state. During the normal course of development, Chd1 gene expression is reduced, the chromatin becomes more compact, and the embryonic cells start on their path to specialization. Much of the molecular machinery for chromatin remodeling is highly conserved throughout evolution and is similar in yeast, *Drosophila*, and mammals, spanning 1 billion years of evolution. If a system works well, why change it?

Gene methylation is another important mechanism whereby proteins keep their regulatory grip on the genes, although in this case it has become favored more recently in evolutionary history, in vertebrates and plants. This process involves enzymes (a class of proteins, as mentioned above) that transfer a chemical called a "methyl group" to cytosine nucleotides in the DNA, especially in the promoter regions. Lots of methylation is associated with silencing of a gene because transcription factors are unable to bind to the promoter to activate the gene, whereas removal of the methyl groups leads to active gene transcription.

Denis Noble has likened the way that genomes build and organize bodies to the way that a large pipe organ is played.[3] Each of the hundreds of pipes contains unique sets of important musical information, but without the proteins, RNA, ribosomes, and all the rest of the cellular machinery needed to play the organ—to extract and use the information from the pipes—nothing happens. An organ needs someone to play it and sits silent until the right musician comes along. There has to be finely tuned coordination between all the hundreds of organ stops, pedals, keys, and pipes for the right music to emerge. And it's not identical music that emerges each time the same piece is played; each organist brings to the piece a particular interpretation. With genomes, too, their expression as a system is intimately bound up with the environmental context of the body that is being built and regulated.

Building *Drosophila* Bodies

Tracking the regulation of the thousands of genes involved in early human embryonic development would take a complete encyclopedia, although as it happens only a small portion of all the events have as yet been elucidated at the molecular genetic level. But in any case, the principles of how genes build bodies are what concern us here, not the detail. Fortunately, once again, *Drosophila* flies to our rescue to provide some stunning examples of the ways that different types of gene are involved.

The *Drosophila* fly has a head, a thorax, and an abdomen, as Figure 3.1 illustrates. The thorax consists of three segments: the first segment carries a pair of legs, the second a pair of wings and a pair of legs, and the third a pair of legs plus what are known as the "halteres." The halteres are a pair of rudimentary wings that help insects maintain stability during flight. The abdomen contains eleven segments.

The big challenge for a very early embryo is to establish its general layout and overall orientation, known as "egg polarity." The *Drosophila* embryo starts by sorting out its front/back and top/underside. Then the number and orientation of the body segments are arranged, followed by the assignment of functions to each segment. A different set of genes organizes each of these stages.

The workings of the "polarity genes" illustrate some important general principles of the role of genes in development. The genes

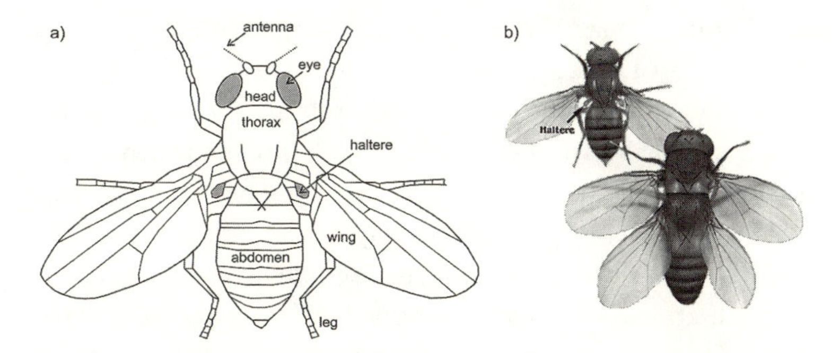

FIGURE 3.1. a) The fruit fly *Drosophila melanogaster*, one of the first and perhaps the most widely studied model organisms in genetics. b) Illustrations of two fruit flies, one wild-type, or normal, fly and one *ultrabithorax* mutant fly. Note that the halteres in the mutant have been replaced by wings. Reprinted by permission from The Nobel Foundation. http://nobelprize.org/nobel_prizes/medicine/laureates/1995/illpres/illpres.html

are already transcribed into the equivalent mRNAs during the formation of the mother fly's eggs, so are all ready for action in the cytoplasm of the egg once fertilization occurs, at which point the mRNAs are translated into their protein products. So it is the female's genes that instruct the new embryo on its basic shape. When it comes to early development of the next generation, the female is definitely in charge.

The polarity gene products work by setting up gradients of "morphogens" along which cell development occurs in either a front/back or up/down orientation. The morphogens are arrayed in a kind of chemical matrix that provides the embryo with precise information about where cells should be in three-dimensional space. At least twelve different genes are involved in the up/down orientation, including *dorsal*, one of those genes already expressed as mRNA in the unfertilized egg, ready to be pulled straight off the shelf, as it were, when its moment comes.

As soon as fertilization occurs, *dorsal* mRNA is translated, and the morphogen is then split between the nucleus and the cytoplasm. The dorsal protein present in the cytoplasm destines that part to become the up side (dorsal side) of the fly, whereas the dorsal protein found inside the nucleus, which has moved to one side of the fertilized egg, destines that side to become the down or underside of the fly. The dorsal protein acts as a transcription factor that switches on a whole cascade of other genes, which in turn lead to body building.

Unlike the polarity genes, the twenty-five or more genes in *Drosophila* responsible for segmentation are only transcribed following fertilization. They become switched on by the products of the polarity genes according to the cascade principle just described, and carry out their actions on the developing embryo according to the further principle of divide and conquer. Mutations (involving changes of nucleotide sequence or even total ablation) in these master control genes can lead to the loss of multiple segments in the eventual fly, making for a pretty unhappy fly.

Most famous of all are the nine Hox genes that define which of the fly's segments does what. Eyes normally arise only on the head segment, whereas legs only appear on the thoracic segments. Hox genes are part of a big class of genes known as homeotic genes, first identified by William Bateson, inventor of the term "genetics," whom we already encountered back in chapter 1. Bateson noticed in his botanical studies published in 1894 that different parts of plants, such as the stamens, occasionally grew where the petals ought to be growing. Such changes were called "homeotic" because they involve the change of one thing into something else (from the Greek *homoiōsis,* meaning a "resemblance").

The Hox genes encode transcription factors that together regulate hundreds of other genes, either by switching them on or by silencing them, again well illustrating the cascade principle.[4] Their role as master control genes can readily be shown by mutating a single Hox gene. For example, mutations in a Hox gene called *antennapedia* cause legs to develop on the head of a fly instead of the antenna. Another well-known example is the mutation of *ultrabithorax,* the Hox gene responsible for placing a pair of legs and the halteres on the third thoracic segment (see Figure 3.1b). When the gene is mutated so its protein is no longer produced properly, this third segment now develops what rightfully belongs to the segment just in front: a pair of legs plus a pair of fully formed wings. Such *ultrabithorax* gene mutations can occur in flies in the wild, which is what initially drew the attention of geneticists to this amazing family of executive genes.

The Hox genes provide the cells in the different segments with a kind of GPS navigator so that they know where they are and what they are supposed to do—except this is a GPS system that uses chemical signals rather than radio waves. Mutations in the Hox genes are like feeding the GPS system with wrong addresses so the cells start making legs, but at entirely the wrong address. It reminds you of those real-life stories about drivers who trusted the voice of their outdated GPS instructor too much and ended up turning left into a river or right into a building site.

The Hox gene–encoded transcription factors each contain a region sixty amino acids long that binds to the promoters that switch on specific genes in the activation cascade. The part of the Hox gene that encodes this stretch is called a "homeobox." These characteristic nucleotide sequences are found in all members of the Hox gene family, as well as in some of the other genes involved in building the fly's segmented structure.

Drosophila melanoster has only four pairs of chromosomes, and remarkably the Hox genes line up on the third chromosome in the same order as the segments line up in the fly. So the genes responsible for the head are found up at one end, with those responsible for the rear down at the other end. Quite why they should be arranged in this way is not fully understood, but it certainly gives a very neat and orderly appearance to the system.

Once the Hox genes had been discovered in *Drosophila,* a search was immediately made for similar genes in other organisms, and they have been found in all animals, except sponges, as well as in fungi and plants. Reach down and feel your ribs (hopefully you can feel them). The segmental pattern indicated by your ribs is generated by your Hox genes. In vertebrates like us, however, the Hox gene system is a bit more complex, and there are usually four clusters of Hox genes on the chromosomes, each cluster containing nine to eleven genes. Some but not all vertebrates have the same sequential arrangement of genes along the chromosomes, and the sequence may be correlated to the timing of development as well. The Hox genes that encode the head are up one end of the cluster and are switched on first, and then the others are activated in sequence, until finally the Hox genes at the other end light up last. It's like playing a wooden flute starting from the top, with the first hole playing the first segment and the bottom hole playing the last segment. Development is literally a question of going head first. If a Hox gene is moved (by genetic engineering, see chapter 11) to a different position in the chromosomal gene cluster, then the timing of its expression is altered according to its new location, suggesting that the location determines the timing.

Fine-tuning of the Hox genes is regulated by the RNA "e-mail system" introduced in the previous chapter, such as the two microRNAs that inhibit expression of the *ultrabithorax* gene.[5] Their overexpression causes a classic homeotic phenotype in the *Drosophila,* which now has wings growing in the place of halteres, just as if the *ultrabithorax* gene itself were mutated. Many microRNAs are encoded strategically close to Hox genes on the chromosomes, ready for regulatory action, and can even be read off the DNA in both sense and antisense directions, both microRNA transcripts serving to regulate expression of a Hox gene.[6]

The long-suffering *Drosophila* provides us with wonderful insights into the giant pipe organ of gene regulation complexity. All the pipes have to play in coordination, otherwise confusion results and the organism no longer develops properly. The sequence and timing of the playing of each note are particularly important. One discordant note and the organism no longer turns out the way it should be—each protein-coding gene, noncoding RNA, microRNA, or whatever has to be exactly right. The system as a whole regulates the development, but each part of the system has to pull its weight. Otherwise, the rest ceases to function properly. "Cooperation and interdependence" is the name of the game. The example that follows shows how these same genetic principles work out at the level of insect social behavior.

The Genetic Regulation of Insect Behavior

Genes can exert powerful effects on the behavior and interactions of whole societies of animals, as seen most clearly in the social insects. Such insect societies also illustrate very nicely the constant flow of information and interaction between genes and environment. Drawing a sharp demarcation line between them is not possible; they are invariably integrated in their impacts.

Insects are incredibly successful as a group of different species. In evolutionary history they were the first creatures to fly, more than 100 million years before reptiles and birds. In many environments they make up 10 to 15 percent of the total animal biomass present in a particular location. In other words, if you gathered up all the animals from an acre of forest and then weighed them all, insects would contribute about 15 percent to the tipping of the scales. In some places they make up more than 50 percent of the animal biomass. No fewer than 850,000 insect species have been named so far, of which 300,000 are beetles, and estimates are that 80 to 95 percent of insect species have yet to be named and classified.[7]

Social insects—which can engage in tournament-based warfare (honeypot ants), agriculture (fungus-growing ants and termites), and complex symbolic communication systems (honeybees)—are found mainly among the termites, ants, bees, and wasps, and represent the most structured animal societies on earth.[8] The last three of these groups are known collectively as the Hymenoptera (derived from the Greek ὑμήν, *humen*, meaning a "membrane," and πτερόν, *pteron*, meaning a "wing"). The species that live socially are known as "eusocial." Some species of Hymenoptera can form huge colonies containing millions of individuals, each belonging to a caste with specialized duties and functions, generally divided into queens and workers. These colonies are often dubbed "superorganisms," so intricately are the different social roles linked together for the good of the colony.

The queens focus on reproduction whereas the workers are involved in colony maintenance, growth, and defense. A colony of social insects may contain just one or a few queens, but from tens to millions of workers. Queens have huge ovaries and can lay thousands of eggs per day, whereas workers may lack ovaries completely. Labor is further divided among the workers. In ant colonies, for example, the smaller workers care for the brood, whereas the larger workers can specialize as soldiers with large jaws and painful toxins (as anyone who picnics in the wrong spot often finds out).

Poke an ant or termite nest (best not done with wasps or bees), and the resulting teeming mass of workers all look remarkably the same. In fact, even though all the insects under observation possess very similar genomes, the adults vary in size up to five hundred-fold.[9] Imagine if the normal weights of humans ranged from 50 to 25,000 pounds! (The Olympics would certainly take on a different character.) A high level of polyandry and recombination, taken together, explain a lot of the diversity seen among the individuals in the colonies, and both terms need some explanation. "Polyandry" simply refers to the fact that the queens mate with multiple males, so the resulting fertilized eggs reflect the genetic diversity found among the different males.

"Recombination" needs a little longer to explain. It's an important mechanism for generating genetic diversity in all organisms that reproduce sexually. As mentioned in chapter 1, during the formation of the germ cells (in the process known as meiosis), the paired diploid chromosomes are reduced to just one of each pair in each germ cell. But before they are separated, the pair of chromosomes snuggle up to each other and actually exchange ("recombine") little stretches of DNA nucleotide sequences. Remember that one of each pair originally came from the mother and the other from the father. In this way the set of single chromosomes (one to twenty-three in the human) that end up in each germ cell of the individual is not quite the same as either one's mother's or father's, truly a fancy way of generating greater genetic diversity.

The difference in the Hymenoptera is that they seem to be the ultimate enthusiasts for recombination of this kind.[10] The honeybees in particular take the prize for the highest recombination rates yet measured for any organism that contains cells with a nucleus, having rates three-fold higher or more than the vertebrates. Why should that be? One reason is that it helps the genetically different offspring to escape the ravages of parasites. Huge social insect communities are particularly prone to parasitic attack because of their communal lifestyle. Everyone who works every day alongside lots of other people, perhaps also dropping off their children at the day-school or nursery on their way to work, knows exactly how the flu spreads among children rather rapidly as they play, and then outward through the parents to the communal workplace. For the social insects the situation is even worse as they're absolutely crammed together, and very susceptible to attack by parasites such as *Wolbachia* that can spread easily in such crowded conditions. But different sets of genes can give greater resistance to parasitic infection, providing some measure of protection,[11] just as they can in other species such as humans.

Eggs of the female Hymenoptera develop into either queens or workers, and environment can have a strong influence here. Female honeybee larvae artificially fed a rich diet develop into queens rather than workers. The palace beckons during times of prosperity. Royal jelly, produced by glands in the head of adult worker bees, is an important component of the food used in cooperative brood care, and a key factor in caste differentiation. But there is also a strong genetic influence in maintaining the right ratio of workers and queens. Too few workers, and who would maintain the brood and defend the colony? Too few queens, and who would carry out all that strenuous reproduction? Genes and environment together perform an integrated eusocial ballet.

Female ants can develop into either winged queens or wingless workers, whereas the males who do the mating are winged. The queens start life with wings, but then shed them following mating once they start building a new colony. The set of genes responsible for wing development are very similar to those used by *Drosophila* fruit flies—no surprise there. Somewhat more surprising is that winglessness in workers is brought about by subverting the "let's-make-wings" genetic instructions at many different steps in different ant species. But the net result is the same: no wings for workers. Wings get in the way when you're scurrying around the ground hunting for food.

The way that genes are intimately involved in the social structure of eusocial insects is well illustrated by the case of the fire ants, so-called because of vicious stings that leave a fiery painful sensation in the skin when bitten, stings used to attack and kill smaller prey. Fire ant nests are made as dome-shaped mounds in open spaces and the ants (*Solepnosis invicta*) produce two types of queen, heavyweights and lightweights. The heavy queens are the Queen Cleopatras of the insect world and head up colonies all alone. The lighter-weight queens establish multiqueen colonies, with anywhere between two and hundreds of queens per colony. Which type of queen is produced is controlled by variant forms of a gene that encodes general protein–9 (Gp-9), a bit of a boring name but an important protein for ants. How genes like Gp-9 vary is explained more in the next chapter, but the main point here is that there are two different variants of the same gene, which are called B and b. These variant types of the same gene are known as "alleles," a bit of jargon that is worth knowing as it crops up so often in genetics. The term "allele" simply refers to the duplicate versions of the same gene on the two paired chromosomes that may be identical or vary slightly in their nucleotide sequence.

Since there are two copies of each gene in each (diploid) cell, and remembering what Mendel discovered, we can see that three combinations of these variant alleles are possible: BB, bb, or bB. The bb combination is lethal, so not very helpful. Workers who have a bB combination accept queens who also carry b, but kill queens who have the BB alleles. And in colonies ruled by a single Queen Cleopatra ant, BB workers do not tolerate any other queens, regardless of which alleles they carry.

How can variant forms of the Gp-9 gene exert such dramatic effects? The answer seems to be that the Gp-9 protein is involved in the pheromone recognition system among ants. As described in chapter 2, pheromones are chemicals given off by animals and plants that are recognized ("smelled") by members of the same species, or in some cases, of other species. Bees give off pheromones from glands distributed all over their bodies. In a colony built on caste, it's like having the right built-in transmitter to give off password radio signals to open doors in a building with tight security. Even if you're royalty, while flashing your bB signal will open all the right doors, if you give off a BB signal by mistake, then regicide will be your imminent fate.

Genetic influences are also present in the finer distinctions of who does exactly what for the colony, giving different members of the same caste distinct jobs. Some genes influence whether worker honeybees collect nectar or pollen. Other genes in ants direct which worker ants guard the entrance and which go out foraging. Nobody wonders what to do next. But evolution is not fast. It's taken about 40 million years, for example, for five agricultural systems to appear in fungus-growing ants by the slow process of accumulating different gene variants.[12] By contrast, human agriculture diversified on a massive scale in just a few thousand years by means of language.

Once the individuals in the different insect castes have been established, massive differences occur in the sets of genes that are switched on in one caste rather than another. For example, in honeybees, out of the roughly ten thousand protein-coding genes in their genomes, no less than two thousand are expressed differently in the brains of queens rather than those of the workers. Among the workers, there are also differences, such as those between honey-bees that forage and those involved in "nursing," looking after the young. MicroRNA activities also vary between different bees doing different jobs. Different social roles demand major differences in the information flow going on from the genes to the different cells in the body.

Having mentioned warring ants with fiery bites, and queens that get killed if they don't have the correct genetic profile, remember, too, that some ant colonies can become quite tolerant of intruders. The Argentine ant (*Linepithema humile*) provides a nice example.[13] In its native South America, the species defends its colonies against other competing ants' nests. But after its immigration to Europe, this creature adopted a more laid-back approach to life, becoming much more tolerant of visitors from other nests. In fact, in Europe this species has formed two immense supercolonies, one stretching over 6,000 kilometers from the Adriatic Coast of Italy to the Atlantic Coast of Spain—known as the Catalonian supercolony—and the main supercolony, which consists of all the other nests. The main supercolony comprises millions of nests consisting of billions of workers. Individuals from either of these supercolonies visit other nests freely, provided only that they belong to the same supercolony.

Even after the two colonies were established in the laboratory for eighteen months, representative workers from either supercolony would still kill each other, whereas workers from nests of the same supercolony 6,000 kilometers apart would be nice to each other. Aggression never occurred between members of the same supercolony. The temptation to draw here all kinds of parallels with European history is strong, but will be resisted. Although the genetic basis for the establishment of such supercolonies has not yet been worked out, we would not be surprised if once again genes that encode "passport" pheromones were found to play key roles, something unlikely to explain the historical vagaries of the same regions of Europe.

The Mysterious Story of FOXP2

When investigating the role of genes in building bodies, or indeed social insect communities, you can either start with the body—or the community—and its properties, and then work down toward the genes, or you can start bottom-up with genes of unknown function, and then work up to see what role they play. The example that follows is in this latter category.

Now and again the discovery and possible functions of a single human gene are so interesting that the gene gains a media life all of its own. This can have both advantages and disadvantages. Advantages include the opportunity to use a single gene to illustrate both the power and limitation of genetics to provide explanations for things. A disadvantage is that the media can often hype the story so much that interpretation races ahead of the facts, even suggesting that there is a single gene "for" some complex human trait, like language, musical ability, intelligence, or sexual orientation, which is clearly not the case.

Such has been the fate of a gene called FOXP2. The name has not, as you might imagine, anything to do with foxes, nor is it based on the fact that investigators have been "foxed" by its precise function (though that would be quite appropriate), but it has a more prosaic etymology. The acronym stands for "forkhead box P2," so-called because FOXP2 is a member of a large family of genes that encode transcription factors containing a "forkhead domain," a distinctive sequence of about one hundred amino acids that binds to DNA. Remember that the role of transcription factors is to bind to gene promoters and switch them on or off. The fact that FOXP2 is a transcription factor therefore immediately alerts us to its probable role, like the Hox genes, as a key regulator of a whole suite of other genes.

As so often happens, the significance of the gene was first highlighted by its inheritance in a mutant form in a family with a medical condition. The KE family has been studied over three generations; about half the family members suffer from various kinds of linguistic difficulty, whereas the other half do not. The inheritance of this syndrome follows the typical pattern of dominant gene inheritance as illustrated in Figure 1.2b (on page 17). Just by looking at the pattern of inheritance, one can conclude that there must be a single gene defect involved: only one copy of the gene on one of the paired chromosomes (pair number seven in this case) needs to be defective in order for the syndrome to occur.

Affected family members are characterized by defects in processing words according to grammatical rules and the understanding of more complex sentence structure, such as sentences with embedded relative clauses. They are quite unable to form intelligible speech. In addition, they have a relative immobility of the lower face and mouth, particularly the upper lip. Medically such impaired ability to perform the coordinated movements that are required for speech is known as "developmental verbal dyspraxia." The impairment of speech and verbal comprehension due to the inability to use the rules of grammar is known as "dysphasia." In the KE family, the disorder affects both verbal and written expression of language.

In 2001 a research team from Oxford University identified the mutant gene in the KE family as FOXP2,[14] and since that time another large family with the same disorder has been found to have a mutation in the same gene passing through several generations. Furthermore, there are isolated cases of the disorder in which pieces of the Chromosome 7 on which the FOXP2 gene is located are disrupted. Given all these data, there is no doubt that the mutant form of FOXP2 is the cause of the linguistic disability syndrome.

When the discovery of the mutant FOXP2 gene was announced, the media pounced on this as a "gene for language" or, even worse, as a "grammar gene." Had they known a little more about genetics, they might have noticed that FOXP2 encoded a transcription factor that must therefore regulate the actions of other genes, so the only accurate description might be to say that "FOXP2 is a gene that regulates other genes involved in the development of linguistic abilities in humans," although admittedly that doesn't sound nearly so gripping as a headline. Indeed, FOXP2 has many downstream targets.[15] The fact that FOXP2 is not just expressed in the brain, but in the lung, heart, and gut as well, should also immediately flag the idea that this gene has quite a range of functions depending on precisely where and when it's switched on.

The misunderstanding implied in talk of a "gene for language" can be seen using a simple analogy. Imagine that some glitch causes my computer to run defectively the Microsoft Word software used to write this book, so that it automatically inserts an "x" as the third letter of every word typed and, in addition, shifts each space to the right by one space. Thxer exsultw oxuldb exg ixbberish, but I don't think one could therefore conclude that the faulty microcircuit must itself encode all the writing software. However, one could infer that it must be an essential part of a much larger system in which it plays a critical role. So, in like manner, FOXP2 appears to be part of a system, probably composed of the products of hundreds or even thousands of genes, that together provide us with the brain, throat, and face structures that underlie our linguistic abilities. To make the analogy work even better, given that the linguistic disorder mutations in FOXP2 are associated with defects in development, one would need the microcircuit to become faulty in the factory that makes the computer, so the glitch would be present as soon as the computer is switched on for the first time.

But what does the FOXP2 gene actually do? Several different approaches can be used to investigate gene function. One approach is to look at the evolutionary history of the gene to determine what role it plays in other organisms that are more amenable to genetic manipulation. FOXP2 is found very widely distributed in animals, including fish, birds, reptiles such as alligators, and all mammals, such as mice. Alligators don't talk, but birds sing, so there has been considerable interest in seeing whether FOXP2 is involved in birdsong. Evidence suggests that it is.

Song learning has evolved in three different orders of birds. Birdsong and human speech are learned at critical periods and involve similar types of specialized brain areas. Birds that learn songs have as a part of their brains an area known (imaginatively) as Area X, essential for vocal learning, which nonlearning birds do not have. During song learning in zebra finches and other birds, FOXP2 is switched on selectively in such brain regions. When the gene is artificially disrupted, then song imitation is defective.[16] In canaries, FOXP2 expression in specific brain areas varies seasonally, with more expression observed at periods when the song becomes unstable. The expression pattern of FOXP2 in the brains of humans and songbirds is actually quite similar, and the regulation of expression of FOXP2 is likely to be the most critical factor in understanding its role.

FOXP2 has also been implicated in bat echolocation, the process whereby bats navigate in the dark by bouncing sound waves off objects and then picking up the resulting echoes. Remarkably, bats can emit echolocation pulses at rates of up to two hundred sounds per second, interpret the resulting echoes within time windows as short as several milliseconds, and make motor responses such as changes in flight maneuvers during these short time intervals.[17] Comparison of the FOXP2 protein-coding sequence from thirteen different species of bat revealed a striking amount of variation relative to the highly conserved nature of the gene among other mammals, birds, and reptiles. Further work is necessary to demonstrate a direct link between these variants and bat echolocation.

The human FOXP2 gene is distinctly different from those of our nearest relatives. The FOXP2 protein, 715 amino acids long, is identical in the chimpanzee, gorilla, and rhesus monkey. The mouse differs in just one amino acid from these three species, despite having last shared a common ancestor with the chimpanzee about 75 million years ago. We last shared a common ancestor with the chimp a mere 6 million years ago (approximately), but our FOXP2 stands out by having two different amino acids. The Neanderthals, with whom we last shared a common ancestor between 270,000 and 440,000 years ago, also had exactly the same FOXP2 version that we now possess.[18] Both differences occur at a site predicted to be important for function, yet what difference, if any, these small amino acid changes make to FOXP2 function in humans and Neanderthals remains to be seen. In fact, given that the level of FOXP2 protein present during particular stages of development or learning is likely to be most crucial in its function, the most important differences may lie in the DNA sequences that regulate FOXP2 expression.

Apart from looking at evolutionary history, another favorite approach in genetic research is to delete either one or both copies of a gene from a mouse colony by using genetic engineering techniques. In mice lacking both copies of the FOXP2 gene (homozygotes), severe developmental defects occur and the mice die prematurely, although development is only mildly impaired in the heterozygotes (only one gene deleted). One particularly interesting defect is observed: normally when very small baby mice are removed from their mother, they give out ultrasonic vocalizations that are important for mother-baby social interactions, but in mice lacking both, or even only one copy of FOXP2, such vocalizations are much reduced. Once again FOXP2 is linked to vocal communication, although such vocalizations for mice are innate rather than learned, which is different from the situation in birds and humans.

In the case of the affected KE family members, the mutation in FOXP2 involves a single change of one amino acid to another in the forkhead region, but just this tiny change in one out of 715 amino acids is sufficient to disrupt the normal functioning of the protein so that it no longer carries out its proper role as a transcription factor. Another genetic approach popular with researchers has demonstrated this effect experimentally. In this case the specific mutation present in KE family members was introduced into the FOXP2 gene in a colony of mice.[19] Once again the mice showed severe deficiency in ultrasonic vocalization, just like the mice who had no FOXP2 at all.

The resulting developmental defects raise a question that is often difficult to resolve when investigating such disorders in humans. Detailed investigations of the brains of KE family members using various scanning techniques have revealed some selective and quite complex differences when compared with the brains of nonaffected family members. But is it the case that the primary impact of the FOXP2 mutation is on the fine-tuning of the face muscles that are needed for proper speech, and that in turn has had a deleterious effect on the development of specific brain regions? Or is it the other way around: that the primary impact is on cognitive development in language regions of the brain, and so in turn lack of use has been the cause of the muscular abnormalities? The jury is still out, although the most likely answer is "both." The striking molecular actions of FOXP2 in the brain are being unraveled to reveal its actions in the development of brain circuitry.[20] Further studies on this fascinating gene will likely reveal more about the complex web of interactions between genes and environment.

What this chapter has illustrated is that to understand body building, it's quite possible to start with a complex system, like the early stages of development of a new organism, or a system of social organization, and then work toward finding the underlying individual genes involved. Equally it's possible to start with an individual gene and then try and work out its role in the organism, an approach that then leads straight into a complex network of interactions. Either way, the pipe organ as a whole still has to play. No gene is an island.

CHAPTER 4
Why and How Do Genes Vary?

Francis Galton (1822–1911)—a cousin of Charles Darwin who enjoyed reading Shakespeare at the age of six, later coined the term "eugenics," and founded modern statistics—was a passionate measurer of almost everything. At one time he managed to construct a "beauty map" of the British Isles by classifying the girls he passed in the streets as attractive, indifferent, or ugly. It emerged that London ranked highest for beauty, and Aberdeen lowest, surely nothing to do with the fact that Galton lived in London. Galton didn't know about genes, but he did note the consequences of genetic variation.

So far we have generally been treating genes as static units that impart fixed portions of information to cells in a coordinated kind of way. In practice, genes do carry out identifiable tasks that remain stable over many generations, but as we noted in the previous chapter, they also vary slightly in ways that affect how efficient or poor they are in carrying out these tasks. Understanding where this variation comes from is a key to understanding evolution, disease, and all those fascinating differences that help to make us unique as human individuals. The implications for evolution, disease, and human uniqueness are largely left to later chapters. This chapter focuses on how the variation occurs. Much of it is internal, coming from changes that occur within genomes, but some of it is external—information flowing into genomes from outside sources. We look at both types, and Figure 4.1 provides an overview.

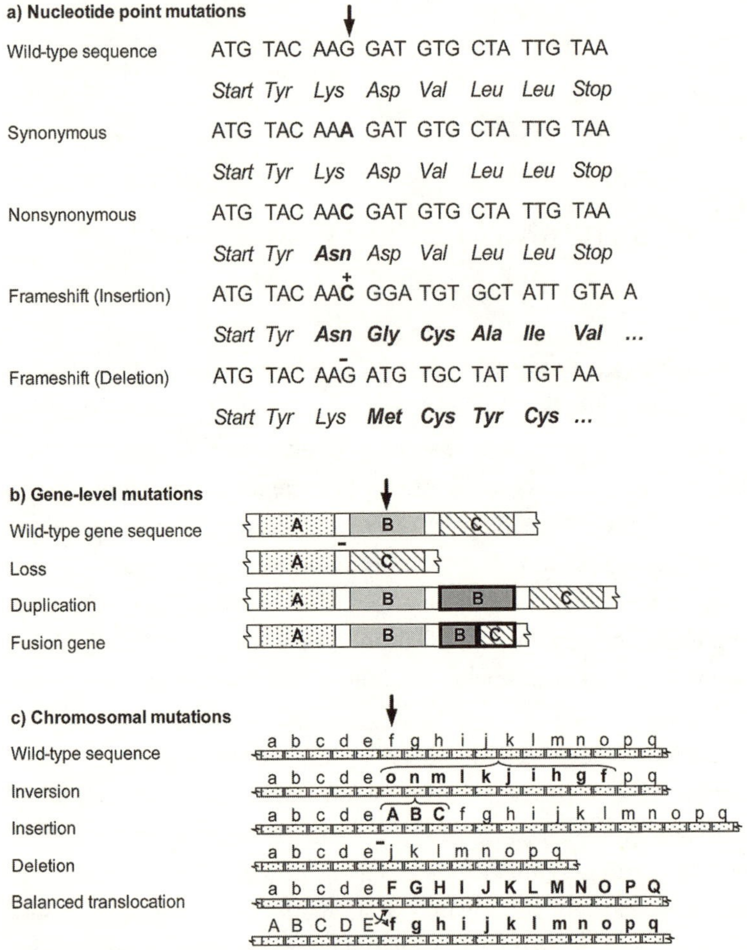

a) Nucleotide point mutations

Wild-type sequence	ATG	TAC	AAG	GAT	GTG	CTA	TTG	TAA	
	Start	*Tyr*	*Lys*	*Asp*	*Val*	*Leu*	*Leu*	*Stop*	
Synonymous	ATG	TAC	AA**A**	GAT	GTG	CTA	TTG	TAA	
	Start	*Tyr*	*Lys*	*Asp*	*Val*	*Leu*	*Leu*	*Stop*	
Nonsynonymous	ATG	TAC	AA**C**	GAT	GTG	CTA	TTG	TAA	
	Start	*Tyr*	***Asn***	*Asp*	*Val*	*Leu*	*Leu*	*Stop*	
Frameshift (Insertion)	ATG	TAC	AA**C**(+)	GGA	TGT	GCT	ATT	GTA	A
	Start	*Tyr*	***Asn***	***Gly***	***Cys***	***Ala***	***Ile***	***Val***	...
Frameshift (Deletion)	ATG	TAC	AAG(−)	ATG	TGC	TAT	TGT	AA	
	Start	*Tyr*	*Lys*	***Met***	***Cys***	***Tyr***	***Cys***	...	

b) Gene-level mutations

Wild-type gene sequence: A B C

Loss: A C

Duplication: A B **B** C

Fusion gene: A B **B C**

c) Chromosomal mutations

Wild-type sequence: a b c d e f g h i j k l m n o p q

Inversion: a b c d e **o n m l k j i h g f** p q

Insertion: a b c d e **A B C** f g h i j k l m n o p q

Deletion: a b c d e j k l m n o p q

Balanced translocation: a b c d e **F G H I J K L M N O P Q**

A B C D E **f g h i j k l m n o p q**

FIGURE 4.1. An overview of mechanisms that generate genetic variation at the a) nucleotide, b) gene, and c) chromosomal levels. For each level an arrow denotes the point at which the mutations are shown to occur, and the mutations and resulting changes are shown in bold. In (a), nucleotide triplets are designated by ATGC and corresponding amino acids are designated by three-letter codes. In (b) and (c), genes are represented by ABCs or abcs.

Different Kinds of Mutation in DNA

We start out looking at the internal kinds of mutation—those that arise within the sequence of DNA itself within a particular cell. Remember that mutations are only passed on to the offspring if they occur in the germ cells—that is, the sperm or the egg. If they occur in the somatic cells—the rest of the cells that make up our bodies—then they are not passed on. Such mutations may, however, lead to diseases such as cancer.

Changing Nucleotides

Every time a cell divides into two "daughter" cells, the DNA in the nucleus has to be replicated to ensure that each new cell has its full complement of chromosomes. The rate of cell division varies hugely between different types of cell in the same organism, and between different organisms. Bacteria such as *E. coli* can divide every twenty minutes in creamy cakes waiting to be eaten in the wedding reception room on a hot summer's day. The longer the speeches, the more bugs—a back-of-the-envelope calculation shows just how many bugs dividing at this rate will be present after twenty-four hours, starting with, let's say, ten.[1] It's a strong argument for short speeches.

In our bodies the rates of cell division are somewhat more modest, but still impressive. When you get the flu, the glands in your neck may swell up as the cells of the immune system, your body's defense system against invaders, go into overdrive to ward off the attack. A class of white blood cell known as T cells, specialized to defend against viruses, divides once every twenty-four hours or so under such conditions. Skin cells are constantly dividing to replace the ones that slough off every day, so they replicate quite fast. Every cell replication involves the accurate copying of the 6.4 billion nucleotides in the DNA of each cell (3.2 billion in each set of chromosomes), an amazing feat that takes place millions of times in our own cells each day—and we don't even think about it.

Every time DNA replicates, an army of proofreading enzymes checks that every G has been copied precisely into a C, and every A into a T, in the complementary daughter strand of DNA. Now and again—in fact, very rarely—the proofreaders miss a copying error, as we all do when we read a text. The result is a mutation, which simply means a change in the sequence. This can happen anywhere in the DNA, in a protein-coding gene, either within an intron or an exon, in an RNA gene, or in that great sea of the 95 percent of our DNA for which the function, if any, is as yet not so clear.

Mutations occur randomly in the genome, which means that they occur without the well-being (or otherwise) of the organism in mind. However, they are not distributed evenly, so are not random in the sense that they occur equally anywhere. For example, studies on bacteria, in which mutation rates can be measured more easily due to their rapid rates of division, have revealed mutational "hot spots" at which mutational change may be up to a thousand-fold higher than average. One reason for this difference is that some nucleotide bases can be chemically modified in a way that makes them more susceptible to change. For example, we have already referred to methylation and the way it regulates gene transcription. When the nucleotide cytosine (C) is methylated to form methyl-cytosine, then it becomes more likely to mutate to a thymine (T). Since a G normally pairs with a C, if the C changes to a T, then a mismatch occurs in the DNA double-helix, and of course the presence of a T rather than a C may introduce gibberish into the gene at that point. In practice, mismatch repair enzymes carry out a quick repair job, but now and again they miss one, allowing the mutation to persist. The importance of this can be seen in human DNA in which only about 1 percent of the nucleotides are methylcytosine, yet about 30 percent of all point mutations are found at these sites.[2]

The consequences of such mutations can range from the benign to the beneficial to the disastrous as far as the organism is concerned. Since the genetic code is degenerate—with more than one triplet codon encoding the same amino acid—some mutations in protein-coding genes make no difference to the resulting protein amino acid sequence at all. Other mutations change the amino acid, sometimes with dramatic consequences for protein function.

Much depends on what difference the mutation makes to the "open reading frame" (ORF), meaning those sections of DNA that the translation machinery can read once the DNA has been transcribed into mRNA. Remember that all mutations in the DNA are converted by copying into mutations in the mRNA. For example, the triplet ATG, encoding methionine, signals the start of translation from the mRNA. So, if ATG at the start of the ORF mutates to ATC, a triplet that encodes another amino acid, then the translation will never get started in the first place, or the translation machinery might keep tracking along the mRNA until it finds another ATG, which it then interprets as a START site rather than methionine, so you only get half a protein. It's as if the entrance to a road is blocked off and disguised with trees so motorists would never know that there was a nice long road there at all—usable but inaccessible. Or if you want a really weak British joke, you could say that normal translation is "switched orf."

At the end of an mRNA molecule, triplet codons signal "STOP translating." One of these is TAA. If this mutates to TAC, then this encodes an amino acid called tyrosine; so instead of stopping at the red light, translation will carry on and a tyrosine will be added to the end of the protein when it shouldn't be there. If some triplets further on encode amino acids, they will be added on as well. Conversely, if we consider the TTA that encodes leucine, and imagine that the middle T mutates to an A to produce TAA, signaling STOP, then translation will abruptly stop at that point in the mRNA. If that happens early on in the sequence, the protein will be severely truncated, so very unlikely to be functional.

Insertions or deletions of one or more nucleotide base-pairs are more common than a change of a single nucleotide. If a nucleotide drops out of the sequence, a deletion, then everything further on becomes gibberish. Take the sentence,

The dog bit the cat and the rat.

Though slightly unusual as biology, at least it makes grammatical sense. But now if we lose a single letter "h" early on in the sentence, we arrive at

Ted ogb itt hec ata ndt her at

which doesn't make any sense at all. The mutation in this case renders the mutation "out of frame"; similar gibberish can result if an extra genetic letter is inserted into the sentence. Once again, metaphors from human language make most sense when speaking of the language of the genes.

As discussed earlier, genes contain introns, promoters, and other regulatory sequences, in addition to the exons that end up represented in the mRNA. Mutations can target any one of these delicate pieces of apparatus. A mutation at a border point between an intron and an exon may mean that an intron gets included in the mRNA by mistake, which is then very confusing for the translation machinery because introns generally don't contain an ORF. Other times a mutation may result in different splicing so that a different set of exons ends up in the protein, which might then have new properties altogether.

Mutations that occur in the regulatory regions can have even more profound effects by changing the timing or level of expression of different proteins. In turn, if the protein being made is a transcription factor like one of the Hox genes, a mutation can cause myriad further effects cascading along the regulatory pathways.

More mutations accumulate in germ-line cells with increasing age, more so for males than for females. This is simply a numbers game. Every time a cell replicates, there is a chance of a mutation occurring, so the more replications, the more the accumulation of mutations. In humans the precursors of sperm cells undergo many more cell divisions than the precursors of eggs, so males contribute roughly six times more mutations to their progeny when compared with females. By the age of 70, a male's sperm cells have undergone around 1400 cell divisions, with a consequent increase in the number of mutations, something to think about for females who choose to reproduce with older men.

Errors made during the proofreading process are not the only way in which mutations can happen. Certain chemicals or radiation can also induce mutations. Hundreds of different chemicals can increase the mutation rate and so are called "mutagens." This is why products such as washing reagents, cosmetics, food preservatives, pharmaceuticals, and the like have to be tested so carefully before going to market. Sometimes the mutagenic effects of chemicals or materials take many years or even decades to appear, as is the case with the mutagens found in cigarette smoke that can lead to lung cancer decades later. Cigarette smoke contains about sixty different mutagens, and smoking is responsible for 90 percent of deaths from lung cancer.

The accumulation of mutations in cancer tissues can be incredibly high, because the cells are proliferating in an uncontrolled way, and also because the DNA repair mechanisms are often defective in cancer cells. Investigation of DNA from a small-cell lung cancer, a type that represents 15 percent of all lung cancers, revealed 22,910 somatic mutations, of which 134 were found in the exons of protein-encoding genes.[3] The cells in the lungs that eventually become cancer cells due to tobacco exposure undergo an average of an estimated one mutation for every fifteen cigarettes smoked. If one now considers all the cells in the lung that are exposed to tobacco, then every single cigarette presumably must cause many mutations in many different cells.

Various forms of ionizing radiation represent other well-known causes of DNA mutations. These can cause mutations in either somatic cells or germ cells, although in practice most mutations will be in somatic cells because they are so much more numerous in the body. The electromagnetic waves shorter than the wavelength of light—such as ultraviolet (UV) light, gamma rays, and X-rays—cause the most damage. For example, the UV light that the body receives in sunlight causes adjacent thymines on the same DNA strand to chemically bind to each other, distorting the DNA structure and sometimes preventing proper replication. People who like to tan with UV lamps in tanning salons or for hours on the beach without sunblock should remember that such activity is a quick way to accumulate more DNA damage in skin cells. UV light is a known risk factor for malignant melanoma. Detailed analysis from the melanoma cells obtained from a patient with skin cancer revealed a staggering 33,345 different mutations.[4] Many of these mutations could be attributed to UV damage because they were caused by the chemical binding of adjacent thymines to each other. In this way UV light leaves behind its own characteristic signature of DNA damage.

Despite all the various risk factors, the encouraging news is that the actual accumulation of new mutations in humans under normal circumstances is rather low, simply because the DNA repair mechanisms function so brilliantly: we all carry around thirty new mutations in our germ-line DNA that were not there in our parents, equivalent to one mutation in each one hundred million nucleotides.[5] Fortunately the great majority of these mutations are benign and do us no harm.

For an example of a really benign mutation (up to a point), consider the outstanding Finnish cross-country skier Eero Mantyranta, who won seven Olympic medals and dominated the sport in the 1960s. Not until the 1990s was it discovered that he has a mutant form of the erythropoietin (EPO) receptor, thereby greatly increasing his red blood cell count.[6] EPO injections are banned in sport today. The whole question of beneficial and deleterious mutations continues to crop up in later chapters as we highlight different aspects of the language of genetics.

Jumping Genes

Another way in which variation can readily enter genomes is through segments of "copy-and-paste" DNA, sometimes known as "jumping genes" or "mobile genetic elements." No less than 46 percent of our own genome consists of this kind of DNA, the bulk of which belongs to a specific class of mobile elements called retrotransposons. More than 3 million examples of retrotransposons have been counted, so jumping genes are extremely abundant. "*Alu*" retrotransposons, for example, consist of three hundred base-pair stretches of DNA and represent the most frequently repeated sequences in the whole human genome, with about 1 million copies accounting for about 10 percent of our entire genome.

Retrotransposons have no known function, except to provide lots of padding for the rest of the genome; these sequences more than any others gave rise to the terms "parasitic DNA" and "junk DNA." However, the insertion sites of retrotransposons can certainly have an impact on the regulation of nearby genes, and occasionally they can change by mutation and acquire new functions.

Why the rather clumsy name "retrotransposon"? This type of gene is copied into mRNA, just like any other gene, but it also has the clever trick (the "retro" bit) of encoding proteins that convert the mRNA back into the DNA sequence again, which is then incorporated permanently into the genome. That's why we can call this a copy-and-paste mechanism. It doesn't take much imagination to realize that if this happened too often, our genomes would soon fill up with so-called parasitic DNA, so cells—not surprisingly—have mechanisms to protect their DNA from too much copying. One kind of protection is methylation of the retrotransposon gene promoters, a mechanism we encountered in chapter 3, which prevents genes from being transcribed.

Incorporation of retrotransposon genes back into the genome occurs more or less randomly, although for chemical reasons the molecular machinery prefers some spots rather than others. Usually the incorporation makes no difference to the functioning of the genome, although occasionally the results can be disastrous—for instance, if the retrotransposon lands smack in the middle of a protein-coding gene, or sufficiently near it to interfere with its regulation. Indeed, more than sixty cases of newly inserted transposons causing disease in humans have been reported.[7] This may be one reason that protein-coding genes are scattered all over the genome and constitute only 1.5 percent of the genome. It's a bit like the strategy of spreading airfields widely all over Britain during the Second World War to make them a more difficult target during bombing raids.

Somatic cells tend to be relatively protected against the copy-and-paste actions of the jumping genes since most of their regulatory DNA sequences remain methylated. They do of course need to keep a whole suite of genes switched on for the daily needs of their particular cell type, which gives the ever opportunistic retrotransposons a chance to do some copy-and-pasting, but it's in the germ cells that retrotransposons get their lucky breaks. Great waves of gene demethylation occur during gamete formation, and again in early embryonic development, so opening a window of opportunity for the copy-and-pasters. Retrotransposons are added to the human gene pool at a frequency of one new insert every ten births.[8] Generally these are benign, but very occasionally devastating genetic diseases are the result—diseases that are passed on to subsequent generations because the events occur in the germ line.[9]

Fascinating studies are also revealing that within a single organism, individual cells can contain different numbers of retrotransposons inserted in different places in their genomes. For example, retrotransposon copy-and-pasting can occur randomly in the early progenitor (parent) cells in the brain of a rat.[10] The result is that the daughter cells then contain different repertoires of retrotransposons, raising the intriguing possibility that brain cells with different properties might be generated in the very same brain, an amazing thought.

Gene Duplication

Normally during DNA replication and cell division, the same number of genes in the parent cell are passed on to the daughter cells. Diploid cells contain pairs of chromosomes, so given one of a particular gene per genome, then each cell will contain just two copies of each gene. But occasionally a segment of a chromosome becomes duplicated and is passed on in the germ-line cells, and this segment may contain one or more genes. Sophisticated methods for detecting these gene duplication events are now available. For example, a comparison of human, bonobo, chimp, gorilla, and orangutan DNA revealed more than one thousand gene duplications that were specific to one of these species.[11] Copy number expansions were particularly pronounced in humans (134 genes specifically duplicated in humans) and include a number of genes thought to be involved in the structure and function of the brain.

Sometimes possessing more copies of the same gene can convey specific advantages. Try chewing for a while on a piece of raw potato; eventually it should start to taste sweet because the amylase present in our saliva breaks down the starch in the potato to form sugar. Individuals can vary between one and ten in their number of amylase genes.[12] Populations that historically have depended on high-starch diets have an average of seven genes all encoding amylase, with consequently higher levels of amylase in their saliva. In a more sinister example, most people have just the normal two copies per cell of a gene encoding a brain protein called alpha-synuclein, but those born with a third copy of the gene are prone to developing Parkinson's disease.

The gene duplication prize at present is held by a giant bacterium named *Epulopiscium fishelsoni*, which means "guest at the banquet of a fish" because it lives in the gut of surgeon fish found in the Red Sea. Most bacteria are only 1/500th of a millimeter across, but *E. fishelsoni* may be up to half of a millimeter across, with a volume up to a million times bigger than a typical bacterium; its huge size meant that not until 1993 was it definitively shown to be a bacterium. Another unusual aspect of this bug is that it nurtures its young inside itself. Most interesting of all for our present discussion is that the bug contains no fewer than eighty-five thousand copies of many if not most of its genes![13] Why this profligacy? One reason may be that bacteria rely on simple diffusion to transport gene products around their single cell. That's fine when there's not very far to go. But in a huge, fat bug like *E. fishelsoni*, that might not be enough. One solution might be to spread thousands of copies of each gene around in the bacterium's single chromosome, and then the gene-encoded proteins won't have very far to diffuse once they've been made.

Large-Scale Mutations

Each chromosome consists of a single double-helix of DNA wrapped up with different proteins. So it might be puzzling that "chromosomal mutations" are considered separately from other types of DNA variation. The difference in this case is one of scale. Chromosomal mutations, sometimes referred to as "structural variations," have historically been those detected visibly by observing stained DNA under a microscope, as Figure 4.2 illustrates. Today a range of technologies make possible the detection of more subtle yet still relatively large-scale changes. Chromosomal mutations can involve many genes—in some cases even hundreds of genes—and generally are more than one thousand DNA base-pairs in length, although this criterion is somewhat arbitrary. The changes may be inherited if they occur in a parent's germ-line cells, but some cases occur de novo, uniquely in that individual. Traditionally the scientists who study such changes are known as cytogeneticists.

Chromosomal mutations occur preferentially on some chromosomes more than others, for reasons not yet fully understood. For example, a preferential involvement of human Chromosomes 22, 7, 21, 3, and 9 takes place in balanced chromosomal variation of all types,[14] so talking about "random variation" leading to chromosomal mutations is not strictly accurate. Variation does not occur with equal probability in any one chromosome.

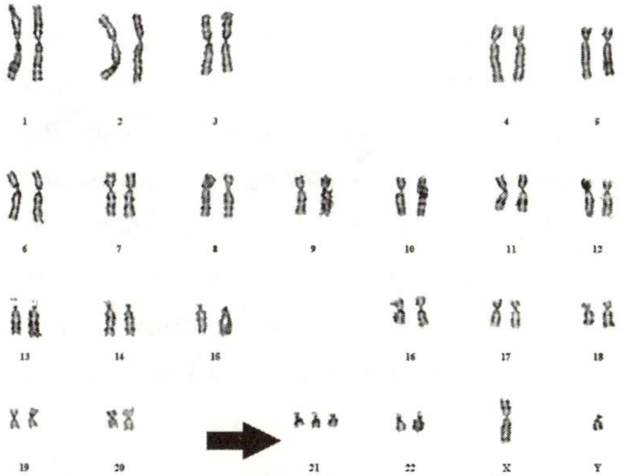

FIGURE 4.2. A karyotype, or "chromosome spread," of a male with Trisomy 21, in which there are three rather than the normal two copies of Chromosome number 21. Trisomy 21 is also known as Down syndrome. Note that chromosomes are numbered according to size, starting from Chromosome 1 as the largest. Reprinted by permission from *Images in Paediatric Cardiology*. Bianca, S. "Non congenital heart disease aspects of Down's syndrome." *Images in Paediatric Cardiology* 13 (2002): 3–11.

Chromosomal mutations come in many shapes and sizes. In chromosomal inversions, a segment of one particular chromosome is reversed end to end, and more than 914 of these have been reported to occur in human chromosomes.[15] In balanced translocations, segments of two different chromosomes are exchanged. The term "balanced" simply means that no net gain or loss of chromosomal material takes place, so many genes would now find themselves in a different location, possibly under different regulatory control. In unbalanced translocations, the same happens except that the exchange of segments now does lead to a net increase or decrease in chromosomal material.

Sometimes the fragmentation point occurs right in the middle of a gene on both the involved chromosomes, so two sections of two different genes are joined together to form a completely new fusion protein. On occasion these fusion proteins lead to cancer, because one half of the gene causes a hyperactivation of the other half, so the gene goes into overdrive and causes cells to proliferate out of control.[16] On other occasions the new fusion protein may have other, more beneficial effects.

Occasionally an extra chromosome is included in a gamete, giving rise to a trisomy in a zygote. For example, as Figure 4.2 illustrates, the presence of three copies of Chromosome 21 ("Trisomy 21") during fetal development gives rise to Down syndrome, characterized by significant mental retardation in severe cases, much milder in others. In other types of chromosomal mutation, a highly specific deletion of relatively small segments can have profound consequences. For example, deletion of 740,000 base-pairs of a specific region on human Chromosome 16 leads to a heritable form of obesity.[17]

Taken overall, large-scale chromosomal mutations can result in conditions such as recurrent miscarriages, infertile males, mental retardation, and congenital abnormalities, although in other cases they make no difference at all.[18] Conversely, and more rarely, they can lead to some significant advantages to an organism. It all depends, of course, on what particular genes are involved in the chromosomal variations. The significance of each card in a game of bridge is determined by the company it keeps in the particular set that you receive.

External Information Flow to the Genome

So far, all the genetic variation discussed refers to that which arises from internal changes taking place within the genome. But genetic variation can also occur by means of imports, and we now consider three examples: retroviral insertions, lateral gene transfer, and those rare occasions when whole packages of DNA are imported inside organelles. As with all genetic variation, the consequences may be positive, deleterious, or neutral for the organism concerned. When we look at these external sources of variation, we will also discuss our evolutionary history back to our most ancient ancestors, since the two topics are so intertwined.

Retroviral Insertions

We normally associate viruses with contracting the flu or other unwelcome diseases. But not all viruses are harmful to us, and they represent a huge pool of genetic information. Viruses consist of some DNA or RNA and a few proteins packaged together to infect host cells where they live as parasites, using the host cells' own molecular production machinery to churn out more copies of themselves.

Viruses are everywhere—two thousand meters down below the surface of the earth, in the sands of the Sahara Desert, and in icy lakes. An estimated 10^{31} viral particles live on the planet,[19] an astronomically huge number (especially considering that there are roughly 10^{22} stars in the universe). A kilogram of marine muck was found to contain up to a million genetically variant viruses. Our own guts may contain as many as twelve hundred different viruses.

An estimated 10^{24} new viruses are created every second somewhere in the world. The vast majority of these die immediately as they are completely unsuccessful in infecting host cells (bacteria in most cases). But the number generated is so vast that fairly frequently their rapid mutation rates ensure that new genes are formed, which may eventually be incorporated into the genomes of other organisms. The huge number of viruses in the world may therefore be viewed as a giant gene production factory, generating a constant stream of new information, some of which other genomes take up and adapt for use.

Retroviruses have the particular ability of incorporating their genetic information encoded in RNA into their host genomes as DNA. A familiar example of a retrovirus is HIV, which causes AIDS by incorporating its lethal genetic information into the body's T cells, the white blood cells that defend us against attack by pathogens. But in many cases retroviruses incorporate their DNA message into genomes without any harm to the individual, and in some cases this infection occurs in germ-line cells, meaning that the inserted DNA becomes a permanent part of the genome of all the descendants of that single individual germ cell. In fact, roughly 8 percent of our own genomes comes from DNA copies of RNA-based viruses that have incorporated themselves into the primate germ line. Each of these long-resident DNA sequences was originally eight thousand to eleven thousand base-pairs long, and they are known as human endogenous retroviruses (HERV).[20] In other words, at some particular time in our evolutionary past, a particular retrovirus has incorporated its DNA message into the germ-line cells of our ancestors, and that sequence has been faithfully replicated over the intervening millions of years. This amazing thought is useful for biologists, as we consider further below, because these retroviral "visiting cards" can be very informative in working out evolutionary lineages.

Retroviral insertions can also provide a valuable resource of new genetic information. Although such DNA sequences for the most part exist as noncoding regions and are initially neutral—neither deleterious nor beneficial for the recipient—with time they can vary due to mutation. In rare instances the sequences can become functional genes, their heritage recognizable by their distinctive retroviral genetic fingerprint. In a few cases, the retroviral insertions include ORFs right from the beginning and have potentially useful functions.[21] For example, bacteria have derived quite a repertoire of useful genes from their infection by viruses, including toxins that they use to kill their competitors.

In the case of the human genome, two genes of retroviral origin, known as syncytin-1 and syncytin-2, are important for normal development of the placenta, as they also are in mice.[22] Both originate from retroviral genes that encode "coat proteins" (the outer proteins of the retroviral particle) involved in cell to cell fusion. For the virus, such proteins are important in causing its fusion with cell membranes and thereby allowing it to enter and infect cells. For the human placenta the genes have been hijacked from the retrovirus for similar kinds of function, but this time in order to play key roles in the fusion of the placental cells with those of the uterine wall, thereby increasing the transmission of oxygen and nutrients between the mother and the developing fetus. As a bonus, the syncytin-1 protein contains a region that may be involved in preventing the mother's immune system from rejecting the fetal tissues.

Having a protein around that causes cells to fuse together is a really useful acquisition when found somewhere like a developing placenta, which indeed needs to do plenty of fusing, in this case with the wall of the uterus. But it's equally easy to see how such proteins could cause real havoc if expressed in the wrong place—rather like squeezing a tube of Super Glue out among the tissues of your body. Fortunately DNA methylation once again comes to the rescue, and it turns out that the syncytin-1 gene is heavily methylated in most nonplacental tissues, ensuring that its functions are focused on the needs of the developing placenta.

The value of such retroviral relics for evolutionary studies is revisited in chapter 7. For the moment, it is worth noting that the retroviral sequence that gave rise to syncytin-1 is present in the most recent common ancestor of the old-world monkeys and the hominoids, but the gene is inactive in the old-world monkeys. Further analysis has shown that the original had to undergo a number of mutational changes before it was fit for purpose and ready for use as a critical player in placental development.[23] So retroviral insertions very occasionally bestow benefits upon their recipient, sometimes without any further modification, but perhaps more often only after some further adaptation.[24]

Lateral Gene Transfer

So far we have viewed the transmission of genetic information as taking place only by vertical transmission, from parents to offspring. But in fact the first 2 billion years of evolution was dominated by quite a different mechanism—the transfer of genes from one bacterium to another, a process known as lateral gene transfer (or horizontal gene transfer).[25] Some people swap stamps or football cards. Other living organisms swap genes. Microbes, we are learning, are like gene-swapping collectives, to use the evocative phrase of the biologist Carl Woese. At least two-thirds of bacterial genes are thought to have originated or been modified by lateral transfer. In one extensive study of more than half a million genes from 181 different types of bacteria, 80 percent of them showed signs of lateral gene transfer.[26]

The consequences of lateral gene transfer are all too familiar for those who have had the misfortune to suffer infection from a strain of antibiotic-resistant bacteria. Indeed, the emergence of resistant bacteria back in the 1950s first drew the attention of scientists to lateral gene transfer. Methicillin-resistant *Staphylococcus aureus* (MRSA) is perhaps the best-known example, the scourge of hospital wards as it is responsible for the deaths of about eighteen thousand patients every year in U.S. hospitals, more than die of AIDS.[27] The problem arises from the overuse of antibiotics in both agriculture and medicine. Low doses of antibiotics are fed to hogs, cattle, and poultry to make them grow faster. It is estimated that in the United States, 70 percent of all antibiotics are used for healthy livestock, 14 percent for sick livestock, and only 16 percent for sick human patients. In fact, more antibiotics are fed to livestock in North Carolina than to all the sick people in the United States.

Multidrug resistance is readily transferred from benign bacteria to disease-causing bacteria because they contain circular minichromosomes called plasmids in addition to their one main chromosome, which contains most of their genes. Plasmids are tiny, containing a few thousand DNA base-pairs and only a few genes, but they can be present inside bacteria in multiple copies, sometimes hundreds of copies. Genes contained within plasmids are not usually required for the growth and replication of the bacteria itself, but rather are used to enable the bacteria to live more efficiently in its host. Plasmids often contain the genes that bestow drug resistance and readily travel from one bacterium to another.

Bacteria demonstrate extraordinary genetic variation, and each species possesses a unique set of physiological characteristics that help it thrive in a particular ecological niche. The size of bacterial genomes is under dynamic regulatory control: genomes gain in size by picking up new DNA by lateral gene transfer, but genomes also quite readily lose sections of DNA that are in excess of requirements. Bacteria seemingly absorb and discard genes according to need.[28]

A dramatic illustration of the consequences of lateral gene transfer comes from seeing how many bacterial genes are present in our own bodies. There are about ten times more bacterial cells in our bodies than our own cells—100 trillion bacteria compared to around 10 trillion of our own cells. Most of the bacteria live in our gut and skin. It might seem puzzling that our bodies can accommodate such a huge number of bacteria—only possible because they are very small: bacteria are typically 0.5 to 5.0 micrometers across, only about one-tenth the size of our own cells.

A study of a large cohort of European individuals from different countries revealed that the group taken as a whole contained around 1,100 different gut bacterial species, and any one individual contained at least 160 of these species on average.[29] Sequencing of the bacterial genomes involved showed that altogether they contained an amazing 3.3 million different protein-coding genes, around 160 times more than the mere 21,000 present in our own cells! We shouldn't feel bad about carrying around our payload of gut flora: although some may occasionally do us harm, the vast majority do us good. For example, our gut bacteria supply us with vitamins, as well as extra energy from complex sugars and proteins that we cannot digest ourselves. In addition, the growth of the harmless majority can crowd out the pathogenic minority, and also have many positive effects upon our immune systems.

Lateral gene transfer is not just important for bacteria; many examples of it occur in eukaryotes as well. A telltale sign of such an origin is when a gene suddenly appears in an evolutionary lineage without any apparent precursor among its ancestors. If the gene in question is also found in bacteria, then the case for lateral gene transfer is strengthened. Lateral transfer seems most common in the "phagotrophic protists" (protists that engulf their food), although it has been described in a wide range of fungi, plants, and animals.[30]

The name "protista" means "the very first," but this ensemble of about eighty different groups of organisms are not really the first organisms to have evolved, although their evolutionary history does appear to go back some 2 billion years. The protists are eukaryotes, often unicellular, and many of them contain chloroplasts, so they derive their energy from the sun by photosynthesis. A well-known protist for those who like looking down a microscope at a drop of pond water is *Euglena*, and those who have seen "red tides" are familiar with the dinoflagellates that can multiply so rapidly that they turn the water red. For clinicians, better-known protists might be *Trypanosoma brucei*, the cause of African sleeping sickness in humans, and the *Plasmodium* that causes malaria.

That some protists are phagocytic provides the clue as to why they might be particularly prone to lateral gene transfer: their food often contains DNA, which is made physically available to combine with the host protist DNA ("you are what you eat"). But in vertebrates like us, in which the germ-line cells are hidden away in special tissues, the chances of being exposed to foreign "invading" DNA are much smaller. Of course, only if transfer occurs in the germ cells would the imported new DNA information be transmitted to succeeding generations.

The choanoflagellates provide an example of a nonphotosynthetic protist that has gained some really useful genes by import. These are little single-celled critters with a collarlike structure (*choano* = collar) made of microvilli that surrounds a single flagellum used for swimming (hence "choanoflagellate"). Do not despise them for their small size (3 to 10 micrometers in diameter). These organisms are thought to be the nearest relatives to animals (metazoans) among the protists and have supplied our own evolutionary lineage with many of the key genes that we need for our well-being. Like us, other organisms have to cope with stress, and all are supplied with a repertoire of proteins that help in physiological stress protection. The choanoflagellates have received at least four of their stress-related genes from another group of protists called algae.[31]

But top prize for genetic generosity goes to a bacterium called *Wolbachia*.[32] This bacterium specializes in infecting arthropods, including at least 20 percent of insect species, making it possibly the most reproductively successful parasite in the biosphere. *Wolbachia* focuses on infecting the testes and ovaries of these organisms and is therefore well-placed to pass on its DNA to its host by lateral gene transfer. Some species are so dependent upon *Wolbachia* that they cannot reproduce effectively without it, whereas in others *Wolbachia* kills the males or causes them to become feminine. A powerful little bug indeed.

Large segments of the *Wolbachia* genome, which contains about 1 million base-pairs in total, have been found in a wide range of bees, wasps, flies, and nematode worms.[33] These are jumping genes, *Wolbachia*-style. Perhaps the most surprising finding of all was that the complete *Wolbachia* genome is found incorporated within the genome of one species of *Drosophila* fly, and *Wolbachia* genes help flies to become more resistant to attack by certain viruses. It is as if you had taken the full contents of your hard drive and simply copied and pasted it into the hard drive of your neighbor's laptop. As a historical event, your neighbor would have no problem discerning when and where the invasion of new information had come from, and it is a parallel kind of reasoning that makes lateral gene transfer so important in evolution, as we consider further in later chapters.

Organelle Import

At certain times during evolutionary history, some rather special and even bigger influxes of information into the genome have taken place than that generated by "normal" lateral gene transfer. The classic and perhaps most dramatic example of this arose when bacteria that had probably started living symbiotically inside cells (that is, they had found a lifestyle that gave some advantage to both partners) then became permanent residents and developed into the mitochondria and chloroplasts that we see in cells today. The mitochondria we now have are little organelles found in the cytoplasm (region outside the nucleus) of all cells of multicellular animals, functioning as the "power plants" of the cell, using food to generate the energy that cells need to keep going. Chloroplasts are organelles found in plant cells responsible for photosynthesis, also "power plants," but with a different energy source (the sun).

Mitochondria and chloroplasts both contain their own DNA, separate from the main storehouse of DNA found in the nucleus. Our own mitochondrial DNA contains only thirty-seven genes, in comparison with the approximately twenty-one thousand protein-coding genes found in our nuclear DNA. Mitochondrial DNA is inherited only from our mothers because it is the cytoplasm of the egg, with its contents, that is used to make more cells following fertilization. The mitochondria from the sperm is lost during this process. This information is quite useful for geneticists who want to track down ancestry, as we discuss later in the chapter about human evolution. In the present context, the key point is the influx of extra information that came with the incorporation of microorganisms into cells, to later become permanent organelles. The process is a bit like acquisitions and mergers. If a big pharmaceutical company has too few drugs in development, the normal strategy is to buy up a smaller biotech company that has some new technology and promising drug leads—both parties benefiting in the process.

How do we know that mitochondria and chloroplasts came into being through an acquisition and merger exercise during early evolution? The great similarities between these organelles and bacteria convinced scientists. The composition of the organelles' membranes is characteristic of bacteria, but their own little genomes are the real giveaway, with more similarities to their parental bacterial genomes than to their host genomes. In 1998 Siv Andersson in

Sweden found that the mitochondrial genome sequence is particularly close to that of *Rickettsia prowazekii*, the nasty aerobic bacteria that cause typhus. This doesn't mean that our mitochondria originally came from this precise strain of bacteria, but it does mean that they came from something very much like it. The word "aerobic" is the key here, because it refers to those bacteria using oxygen for their metabolism ("anaerobes" live without oxygen). The incorporation of an aerobic bacterium into early cells involved an enormous influx of useful genetic information, bestowing upon them the ability to use oxygen, in turn having a huge impact on evolutionary history. In a similar tour de force, chloroplast DNA has been proved very similar to the DNA found in cyanobacteria, the light-harnessing bacteria that live in oceans and fresh water. Today mitochondria and chloroplasts have both become completely dependent upon many nuclear genes for their construction and replication. They have given up their independence but in the process found a permanent home.

Whether by internal mutation and organizational changes, or by importing novel information from other sources, genomes have become specialized to change. Life is a dynamic, active, ongoing process, and genetic variation provides the raw material for evolution. Just how that works in practice we now consider.

CHAPTER 5
How Genes Rescued Darwinian Evolution

FOR MANY YEARS Charles Darwin sought to understand the mechanism of biological inheritance, but as noted in chapter 1, he never did find out about genes. Fortunately, failing to understand one biological mechanism did not prevent him from discovering another, and he published his theory of evolution in *On the Origin of Species by Means of Natural Selection* (1859). "Natural selection" is shorthand for differential reproductive success among organisms based on their heritable traits, as explained in greater detail below. It is essentially a simple idea, if not intuitive. After hearing Darwin's new theory, Thomas Henry Huxley exclaimed, "How extremely stupid not to have thought of that!"

Darwin may have felt the same way if someone had been able to explain genetics to him. Surely he would have been delighted when scientists began to do what he had wanted to do but could not—to combine genetics with his theory of natural selection. The resulting "neo-Darwinian synthesis," as it is sometimes called, is what makes contemporary evolutionary theory so powerful.

When scientists use the word "theory" it has a technical meaning different from its everyday usage to express doubt, such as when we say, "Oh well, it's only a theory." A "theory" in science is much more like a map that incorporates and renders coherent a broad array of data. When scientists talk about the "theory of gravity" or the "theory of relativity," they are not expressing any doubt that things in general fall downward, nor that the relative motion of observers changes what they observe. In a similar manner, the theory of evolution acts like a reliable, conceptual map to render coherent a huge array of disparate data, including data derived from genetics, fossils, dating, anatomy, physiology, and the geographical distribution of species.

The aim here is not to present the evidence for evolution, which has been done very effectively in many publications,[1] but rather to focus on the function of genes in the evolutionary process. Some further historical background is useful, picking up the story from chapter 1. The twists and turns that characterize the history of genetics in the context of evolution remind us once again that progress in science is far from linear.

The Decline and Revival of Natural Selection

Darwin died in 1882 and was buried in Westminster Abbey with great pomp as a British scientific hero. But ironically, for the following fifty years his theory of natural selection actually declined in popularity, and by 1900 some biologists were talking about the demise of Darwinism. In 1903 the German botanist Eberhard Dennert proclaimed, "We are now standing by the death-bed of Darwinism, and making ready to send the friends of the patient a little money to ensure a decent burial of the remains."

Evolution as an idea remained immensely widespread and popular, but *how* evolution actually happened was widely disputed. The significance of Mendel's key results remained unknown. Lamarckian evolution, the inheritance of acquired characteristics, retained its appeal, because the sudden jumps that were observed in the fossil record seemed better explained in this way.

Even great enthusiasts for evolution, such as Darwin's friend Thomas Henry Huxley, never really accepted slow-acting, incremental natural selection as the mechanism for evolution, much preferring the idea of big jumps or so-called saltations. The secular

Huxley was also suspicious of the role of chance in generating variant phenotypes of organisms upon which natural selection then acted. For Huxley, chance sounded like an opening for God's special creation, whereas he wanted to see evolution emerging out of natural scientific laws. It is ironic that in his day Huxley resisted the idea of chance because he thought that it had theological overtones, whereas creationists today resist the idea of chance because they think that it has atheistic overtones. People often interpret essentially the same data in quite different ways depending on their political, economic, and cultural contexts.

The great Victorian idea of Progress also seemed to fit better with Lamarckian ideas. Surely it is more rational, so the argument went, that the useful things that animals learn during their lifetimes should be passed on to their offspring. Why waste what you've learned? Let it benefit a future generation. This again illustrates the influence of one's own political or social ideologies upon the interpretation of data. Letting the data speak for itself is not as straightforward as it may seem.

The rediscovery and extension of Mendel's results around 1900 were described in chapter 1. You might have imagined that once the Mendelian laws of inheritance had been rediscovered, then obviously they would be brought together with the idea of natural selection to generate very quickly the kind of theory of evolution that we have today. But that didn't happen right away. During the early decades of the twentieth century, Mendelism, as the pattern of inheritance that Mendel discovered became known, was actually seen as a *rival* to the theory of natural selection. How come?

The answer is that the particulate concept of inheritance readily lent itself to the notion that changes in evolution happen rather suddenly. For example, the botanist Hugo de Vries made extensive studies of the evening primrose, observing that new, differently colored varieties sprouted seemingly at random. His so-called mutation theory became the most popular theory of evolution in the early decades of the twentieth century. At that time, the term "mutation" primarily referred to the apparently sudden appearance of a distinct form. Only much later did the word take on its contemporary meaning as described in the previous chapter: a physical change in DNA that may cause a change in phenotype.

The early twentieth century therefore saw Mendel's results on particulate inheritance identified with the idea of mutations, perceived as saltations (sudden jumps), such that speciation itself was thought to happen abruptly. To some biologists this made Darwinian natural selection appear superfluous. If new varieties or mutations could come about suddenly, then why did you really need natural selection? Other biologists, and de Vries was one of them, retained a negative role for natural selection in eliminating the unfit mutational varieties that arose, but they didn't credit natural selection with the power to establish new varieties through incremental adaptation. Instead it was thought that species occasionally went through rapid bouts of mutation in which they generated a whole selection of new varieties, which also explained, so it was thought, the gaps in the fossil record.

Now what all this shows is that it's not a good idea to base general conclusions in biology on the study of just one or a few species. By 1920 it became clear that the evening primrose that de Vries had been studying for so long was a complex hybrid, and his apparently new forms of primrose were not new examples of mutation at all, but simply recombinations of existing characteristics.

New insights came from investigations on quite a different organism. Thomas Hunt Morgan's groundbreaking work on *Drosophila* during the first few decades of the twentieth century, showing that genes were strung out on chromosomes "like beads on a string," was described in chapter 1. Morgan was a laboratory-based experimentalist who initially saw little role for natural selection and was greatly impressed by de Vries' mutations. He decided to see if he could produce the same type of saltationist mutations in his flies, but what he found were small but definite variations that were inherited in a Mendelian fashion. Through these observations and many conversations with his students and colleagues, Morgan finally came to accept that Darwin's theory of natural selection could account for the origin of species, although he always emphasized the role of mutation in providing the material upon which selection acts.

Birth of the Neo-Darwinian Synthesis

The next important stage in the development of evolutionary ideas in biology came not from plant breeders, nor from fly breeders—indeed not from the laboratory at all—but from population geneticists and mathematicians. The key question now was, how did evolution actually work in populations of living organisms out in the wild? Three famous figures were associated with this shift in thinking: the mystic British communist J. B. S. Haldane, the Anglican British eugenicist R. A. Fisher, and the American Sewall Wright, the son of first cousins, who was a professor at the University of Chicago.

These three scientists were the first to apply mathematical analysis to the study of genetic variation in a population, and it was the fusion of mathematical genetics with the theory of natural selection that later came to be known as the neo-Darwinian synthesis. Biologists at that time were so unused to mathematical treatments of their subject that Fisher's first paper submitted to the journal of the London Royal Society was turned down because no one could understand it! However, once explained, it was clear that this new approach was very useful. Most notably, the trio used it to identify four main evolutionary factors that affect how gene frequencies in a population change over time: genetic drift, natural selection, mutation, and gene flow.

As we've seen in previous chapters, not all copies of a particular gene are identical, and it's necessary to distinguish among the different variants of a gene when examining how these evolutionary forces shape gene frequencies in a population. As noted in chapter

3, the term "allele" simply refers to different versions of the same gene that may be identical or slightly different, and in chapter 1 we saw how healthy and disease-related alleles of the same gene can have very different effects on an organism. If there's no obvious advantage or disadvantage among the alleles of a given gene, the evolutionary factor most likely to describe their frequencies is genetic drift.

Genetic drift is the change in the relative frequency at which an allele occurs in a population due to random sampling and chance. The alleles in offspring are a random sample of those in the parents, and chance has a role in determining whether a given individual survives and reproduces. As Figure 5.1 illustrates, imagine that you put twenty marbles in a jar to represent twenty organisms in a population. Half of them are black and half white, corresponding to two different gene alleles in the population. The offspring they reproduce for the next generation are represented in another jar. In each new generation the organisms reproduce at random. To represent this reproduction, randomly select any marble from the original jar and deposit it in the second jar. Repeat the process until there are twenty new marbles in the second jar. The second jar then contains a second generation of "offspring," twenty black or white marbles. Unless the second jar contains exactly ten white and ten black marbles, there will have been a purely random shift in the allele frequencies, which will then influence the allele frequencies of the next generation as well. In due course all of the marbles will be entirely white or entirely black, and the other allele will then have been lost.

While drift constantly and randomly reduces genetic variation, natural selection provides another, more selective winnowing force. It acts as a powerful sieve, just as Darwin had always maintained, filtering out those sets of alleles that reduce the fitness of the organism. Fitness here doesn't refer to the consequences of going to the gym regularly, but is rather a shorthand way of expressing reproductive success.[2] Organisms well-fitted to their environment

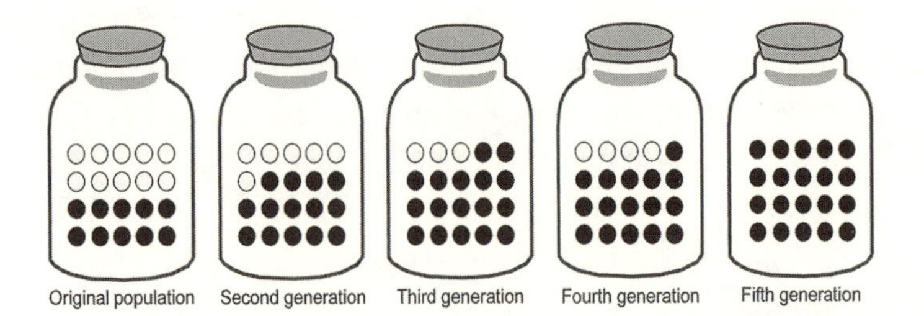

FIGURE 5.1. An illustration of genetic drift. The first jar on the left contains a "marble population" in which two colors of marbles represent two different genes at a 50:50 ratio. In each generation ten marbles are selected at random from the jar representing the previous generation. Given enough "generations," one type of marble will always go to fixation and the other will be lost.

are those that generate plenty of progeny in succeeding generations. Sewall Wright pictured selection acting on an "adaptive landscape" in which mountain peaks represent well-adapted sets of genetic variants, or genomes, and valleys represent poorly adapted genomes (Figure 5.2).

If drift and natural selection were the only forces responsible for shaping the genetic variation of a population, variation would largely disappear. Mutation and gene flow thus serve to add genetic variation to a population. In the context of population genetics, "mutation" in the 1920s and 1930s came to mean a discrete change in a gene, resulting in a distinct allele, even though chemically these changes weren't understood at the time because DNA wasn't yet understood.

Gene flow simply refers to the transfer of alleles of genes from one population to another in the same species. Let's imagine that two animal populations have been breeding quite separately on either side of the country. During their time apart they accumulate different sets of allelic variants. They then migrate and mingle, and start interbreeding again quite randomly. The transfer of variant alleles from one population to the other is called gene flow.

So genetics is what rescued Darwinian natural selection from

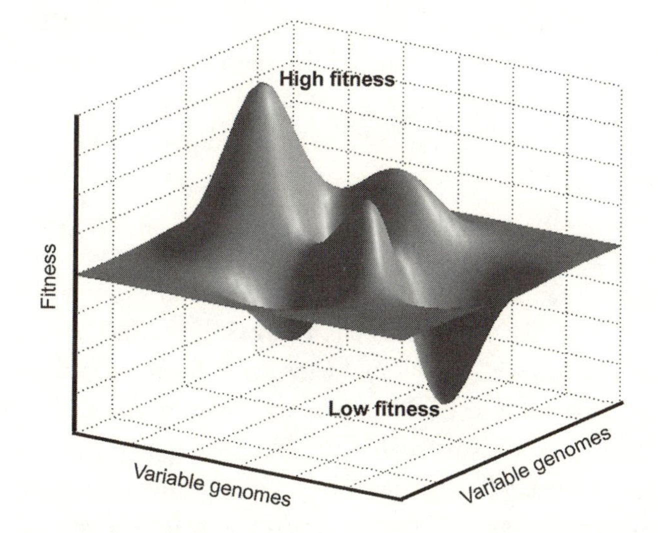

FIGURE 5.2. The adaptive landscape, suggested by Sewall Wright, is a way to visualize the relationship between genes and evolutionary fitness. Peaks on the landscape represent high fitness while valleys represent low fitness. Different combinations of genes yield different degrees of fitness.

oblivion, forging the neo-Darwinian synthesis, which continues to provide a powerfully effective map for explaining the origins of all biological diversity on planet Earth, both now and historically. Big ideas in science often benefit from scientists who are good at communicating the key results to a wider public, and the unusual J. B. S. Haldane did precisely that for the synthesis. Haldane has been described by Stephen Jay Gould as "independent, nasty, brilliant, funny, and totally one of a kind."[3] He learned Mendelian genetics as a boy by breeding guinea pigs and often served as one himself when he helped his father, who was professor of genetics at University College London. In one childhood episode, his father made him recite a long Shakespearean speech in the depths of a mine shaft to demonstrate the effects of rising gases. When the gasping boy finally fell to the floor, he found he could breathe the air there, a lesson that served him well later in the trenches of World War I.

Later Haldane himself quite often experimented using his own body, one time drinking a large quantity of hydrochloric acid to observe its effects on muscle action. I hasten to add that none of these experiments should be repeated by anyone reading this text, but it's perhaps not surprising that the writer Aldous Huxley incorporated Haldane into at least one of his novels as the archetypal eccentric scientist.

More relevant to our immediate topic is Haldane's ten highly mathematical papers published between 1924 and 1934, plus his influential book *The Causes of Evolution* (1932), in which he reestablished a central place for natural selection in the neo-Darwinian synthesis. It's interesting that Haldane comments in his 1932 book, "Criticism of Darwinism has been so thoroughgoing that a few biologists and many laymen regard it as more or less exploded."[4] This statement shows just how far the drift away from Darwinism had gone since 1882. But Haldane's aim was to resurrect Darwinism by showing that continuous, small-scale variation could also have a Mendelian basis, and especially that tiny selection pressures, working in a cumulative manner on such minor variations, could effectively explain evolution.

Haldane was a theoretical biologist who never did much fieldwork, but he did use the famous results of the biologist J. W. Tutt on the peppered moth. Increasing industrialization in Britain had led to a higher proportion of black moths, presumably because they were less visible to predators as they rested on sooty leaves. Haldane calculated that the observed increase of black moths from 1 percent in 1848 to 99 percent in 1898 required only a 50 percent higher survival rate of black moths over speckled ones. But if the increase was solely due to variation without selection, as the early Mendelians tended to argue, then this would require one in five moths to mutate from speckled to black, an obvious impossibility. Although more recent work has improved the quality of the data, Haldane's general point was a sound one.

Other influential biologists followed up in popularizing the new neo-Darwinian synthesis. Julian Huxley, brother of Aldous and grandson of Darwin's great defender, Thomas Henry Huxley, was the author of *Evolution: The Modern Synthesis* (1942), one of the most influential books on evolution in the twentieth century. He carried out famous studies on the Great Crested Grebe and on other birds that mate for life, developing ideas that Darwin himself had originally discussed on the evolution of sexual selection. Like Haldane, Julian Huxley was one of the biologists in the early twentieth century who restored a prominent role to natural selection in the evolutionary narrative.

Other key figures who helped establish the neo-Darwinian synthesis include the Russian, later to become American, Theodosius Dobzhansky, a committed Eastern Orthodox Christian who was a student of Morgan and first used genetics to investigate natural populations of *Drosophila* in the field. The title of one of his popular papers, "Nothing in Biology Makes Sense Except in the Light of Evolution," published in 1973, has become almost a mantra in the field of evolutionary biology. It's interesting to note how three of the great founders of contemporary neo-Darwinian theory—Haldane, Fisher, and Dobzhansky—represent such an interesting range in their own religious commitments. Haldane was the atheist albeit mystic Marxist; Dobzhansky, the Eastern Orthodox; Fisher, a committed Anglican who sometimes preached in his College Chapel in Cambridge—a good example of how scientists of any faith or none can contribute in the scientific enterprise to establish a common theory.

Today the neo-Darwinian synthesis continues to provide a powerfully effective map for explaining the origins of all biological diversity on planet Earth. It is neither a perfect nor a static theory, and the map of life that it provides will surely be adjusted and refined to incorporate new data as it emerges. Yet the evolutionary map, even in its present form, remains one of the most stunningly successful theories in the history of science.

Genetics and Darwinian Evolution Today

This historical introduction should make clear that evolution is currently understood to be the result of two steps or processes, both of which are critical for evolution to occur.

First, there is the generation of genomic diversity by all the many different mechanisms we have reviewed so far. There is the important mechanism of recombination during meiosis, when segments of the paired chromosomes are exchanged during the formation of the sex cells, as described in chapter 3. Sexual reproduction itself leads to massive amounts of variation in a population as the different chromosomes from two parents are mixed and matched to generate different progeny. Genetic novelty might also come from any or all of the mechanisms reviewed in chapter 4: point mutations; insertions and deletions; retrotransposons (the jumping genes); gene duplication and divergence; genome duplication; chromosomal mutations; retroviral insertions; lateral gene transfer; and, very occasionally, the import of a complete external genome from some other organism. And let's not forget gene flow, described above.

The second main step, natural selection, operates to test the variant phenotypes generated by the different genotypes in the workshop of life, so that a population of organisms becomes better suited over time to its environment. It works because the genes of individuals that produce more offspring are better represented in the next generation than the genes of individuals that produce fewer offspring. Thus, a genome that, on balance, aids an organism's survival and reproduction will be "selected for," being more fit, whereas a genome that, on balance, hinders survival and reproduction will be "selected against," being less fit.[5] Selection acts on the individual, which is a product of its many genes and their interactions with each other, some advantageous, some disadvantageous, and some neutral. The key criterion is that of reproductive success: how many progeny representing particular ensembles of genetic variants in their genomes are generated, and how many of these pass their genomes on to subsequent generations?

The term "survival of the fittest" has sometimes been used to describe natural selection, but is not very accurate because survival is not really the main point in this process. Of course, if an animal or plant does not survive then it won't reproduce, but the key point about natural selection is the successful reproduction that allows an individual's genes to be passed on to the next generation.

Evolution is thus a two-step process: generation of diversity followed by rigorous filtering of that genetic diversity in the workshop of life. The great majority of genetic changes, if not neutral, are likely to be maladaptive, so they disappear from the population after some generations—or even immediately if they are lethal. On the other hand, the few beneficial changes that readily pass through the filter of natural selection spread throughout an interbreeding population as they bestow reproductive benefits on their recipients. Alleles that increase fitness in this way are said to become "fixed" in the population once all individuals possess the same allele.

Having surveyed in chapter 3 the various kinds of ways in which genes build bodies, it should be clear that the effects of each allelic variant of a gene are defined by the company it keeps. It takes many different genes to organize development, construct a complex metabolic pathway inside cells, or build a limb or an eye. A variant allele that might bestow benefits when its effects collaborate with ninety-nine other alleles to produce a certain outcome may no longer be beneficial when present along with a slightly different set of ninety-nine alleles.

The metaphor of the selfish gene has been used to present a gene's-eye view of evolution, in which the gene is envisaged as a selfish replicator occupying a "survival machine" that only exists for the gene's benefit to convey it onward to the next generation.[6] The metaphor has some use in that it draws attention to the importance of each variant allele in building organisms, but overall it is somewhat misleading. In reality each gene is dependent on the actions of many other genes. Genomes provide systems for building organisms in a cooperative, interactive way; there is nothing "selfish" about that. And the functioning of one gene may change considerably depending on the presence or absence of variants of other genes. So the metaphor of genes as "cooperators" might be more accurate as a way of describing how they collaborate together in real life to carry out complex, interacting functions.

We can picture the way in which genes cooperate like an orchestra. In a real orchestra a musician playing a particular instrument may flourish if the environment provided by the rest of the orchestra generates the best interactions to play a great symphony, but equally the sound of even a very gifted player is quickly spoiled if the rest of the performers are off-key. So it is with genes: the very same gene can exert a rather different effect in building and running the organism depending on the company that it keeps. Strictly speaking, therefore, an allele cannot be defined as "beneficial" or "deleterious" as if this were its permanent characteristic. It all depends.

Genomic data have also revealed that the notion of a "neutral" allele—that is, a variant gene that supposedly neither increases nor decreases fitness—is also problematic. You might have thought that mutations that cause no change in the amino acid sequence of proteins (because they result in an alternative codon use for the same amino acid, as discussed in chapter 2) would be "neutral" rather than having any influence on selection. After all, exactly the same protein is produced. However, such mutations can in fact make a difference to evolution.[7] One reason is that some tRNAs (which, remember, are specific for only one particular codon) are much more abundant than others, leading to more efficient translation with higher fidelity for one codon than another, even when they encode the same amino acid. Mutations may also lead to differences in mRNA stability and alternative splicing, even when they do not change the amino acid sequence.

The impact of variant alleles may also vary greatly depending on whether they are present in either one or two copies, as illustrated by the diseases discussed in chapter 1. Sometimes the heterozygous condition, in which a mutant allele exists on one of the pair of chromosomes but not the other, is actually beneficial for its carrier, even though the homozygous condition for the same mutant allele is profoundly deleterious. This kind of situation leads to a balance of genetic variants in the population in which a certain proportion of heterozygotes is maintained, due to the advantages bestowed, despite the fact that occasionally heterozygotes will mate, with a one-in-four chance of any one of their progeny having the homozygous deleterious condition.

An actual example of this scenario is provided by an allele that helps protect against malaria. Each year, about 400 million people contract malaria, and 2 million to 3 million die, the majority of whom are children. Malaria is caused by a parasite that infects red blood cells and feeds on hemoglobin, the protein responsible for transporting oxygen in our blood. In parts of the world where malaria is common, such as some African countries, a particular allele of the hemoglobin gene is also common. This "hemoglobin S allele" contains a single-point mutation that makes the hemoglobin protein prone to clumping, which in turn makes the normally disc-shaped red blood cells become elongated and rigid, or sickle-shaped, as Figure 5.3 illustrates. The parasite finds it hard to feed and reproduce in the sickle cells, so the risk of malaria is reduced. A double dose of hemoglobin S ("homozygous") means that the sickled blood cells clump even more, blocking capillaries and preventing oxygen from being carried to the body's tissues. Imagine pushing a bunch of oranges down a pipe with lots of twists and turns, and then try it with bananas—it gets much harder. This homozygous condition is called sickle-cell anemia, and about 80 percent of people who have it die before reproducing. But natural selection doesn't weed the hemoglobin S allele from the population, even though it's very deleterious in the homozygous condition, because it bestows the partial protection against malaria to heterozygotes.

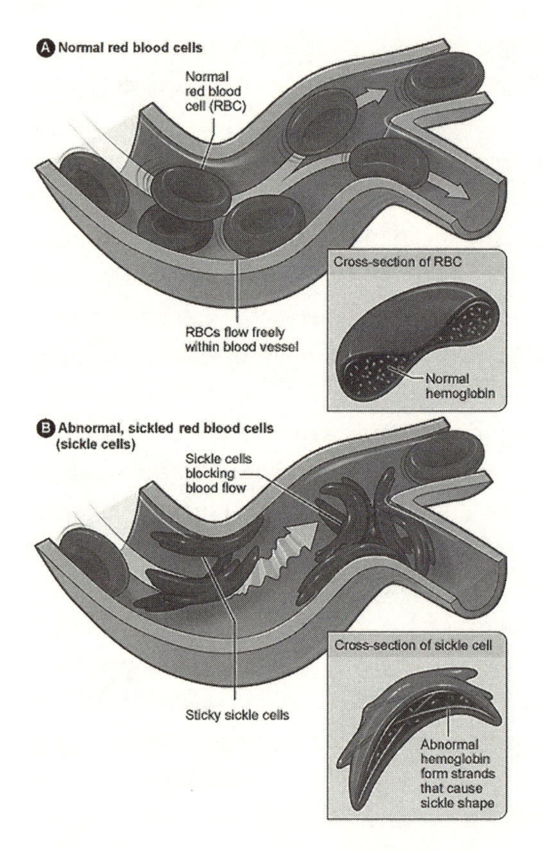

FIGURE 5.3. Normal (A) and sickled (B) red blood cells. Sickle cells are a result of a mutant form of hemoglobin that is less flexible than normal hemoglobin and distorts the shape of the blood cell. One copy of the mutant form of hemoglobin provides some protection against the parasite that causes malaria, but two copies of the mutant hemoglobin gene cause sickle cell disease in which blood circulation is impaired by the profusion of sickle-shaped cells. Reprinted by permission from The National Heart, Lung, and Blood Institute. http://www.nhlbi.nih.gov/health/dci/images/sickle_cell_01.jpg.

In practice, natural selection does not weed out all deleterious alleles, even when they bestow no particular advantage upon the organism in the heterozygous condition, as in the example above. The reason is that deleterious alleles often hitchhike along with a beneficial allele located nearby on the chromosome. We say that alleles are "linked" when they tend to travel together during recombination and get inherited as a "package." Of course, if the net effect of the deleterious alleles outweighs that of the beneficial allele, then the overall fitness of the organism will decrease, and this genotype will tend to be weeded out by natural selection. But if the deleterious allele is relatively mild in its effects, then hitchhiking will be successful.

Whether variant alleles are beneficial depends not only on the genetic company they keep at the genomic level, but also on the environment of the organism that they help to build. Think about the anteater, which needs a long snout to delve down into big anthills and narrow crevices to fish out those delicious little ants, finding his snout getting longer and longer over succeeding generations only so long as its advantages outweigh its disadvantages. Crevices are only so deep, and at a certain point the snout gets so long that it slows down escape from predators. So a kind of equilibrium is reached in which genomes build snouts of just the right length to do the job that needs to be done without being a handicap: anteaters with that optimally useful length of snout will eat lots of ants, flourish, and have loads of offspring, passing on their useful genomes to succeeding generations. Notice that this natural selection process is very different from the Lamarckian idea that the anteater tries hard to get ants just out of reach, thereby lengthening her snout, so passing her long snout on to her offspring. That is not how it happens!

But if the supply of ants suddenly dries up for some reason, an anteater with a long snout may be in trouble. A different selection pressure now begins to operate, and adaptations that enable anteaters to successfully use alternative food sources might develop. The ecological niche defines what kind of adaptations will develop. An anteater's long snout is not going to be of much help to polar bears.

Natural selection is generally a rather slow process: steadily winnowing, shaping, weeding, and revising genomes along with their attendant phenotypes. Biologists in a hurry to find answers turn to organisms that reproduce fast, such as bacteria that can divide every twenty minutes if fed well, and examine their evolution in the laboratory.

Rich Lenski and colleagues, then at the University of California, did exactly this on February 24, 1988. They started growing a series of twelve populations of the bacterium *Escherichia coli,* all derived from a single bacterium and fed using glucose. The evolution of different strains of these bacteria from the original parental cells has now been tracked for a period of more than twenty years.[8] Each day about half a billion new bacteria grow in each flask, involving the replication of the same number of bacterial genomes, and in total about a million mutations occur in each flask as the bacteria divide. Since there are only about 5 million base-pairs in the bacterial genome, this means that every few days virtually the whole genome will be subject to genetic analysis to see whether any of the new mutations might be useful. In practice the vast majority are not, but new mutations occasionally come along that provide some growth advantages.

Every night the bacteria run out of their glucose food source and become dormant, so bacteria that cope best with this changing environment have a big advantage. The next day about 1 percent of the culture from each flask is used to start a new culture with a new supply of glucose. Most of the beneficial mutations that occur provide up to a 10 percent growth advantage, and such mutations spread rapidly through the population as the progeny carrying the mutation have this modest growth advantage. What Lenski found was that the evolution of the different flasks of bacteria, as measured by their growth, developed not in a smooth trajectory but in a series of abrupt jumps as advantageous mutations took over the population. These are not quite the "saltations" that were popular in the early twentieth century, but certainly highlight the fact that even a small change in one or a few genes can make a big difference to the success of the organism.

After more than a decade of subculturing the twelve flasks, something rather extraordinary happened at generation 33,127. One of the cultures "discovered" how to use citrate as a food source, a chemical used to stabilize the pH and so present in all the flasks since the beginning. It was like a population of cats suddenly taking a liking to whisky (assuming an unlimited supply) and gave this population a huge growth advantage as it was no longer dependent upon glucose as a food source. This critical event happened in only one of the twelve flasks and it took more than ten years to show up. Further analysis revealed that the capacity to use citrate could not evolve all in one step, but took three different mutations to achieve. The two "background" mutations had to occur first, and the third critical mutation then enabled the complete ensemble of three mutations to allow the use of citrate, thereby opening up a whole new way of living for the colony. In fact, what has happened is that 99 percent of the colony uses citrate, whereas about 1 percent have become "glucose specialists," stubbornly refusing to forsake their original food source.

This wonderful experiment highlights the need for multiple cooperative alleles to emerge together in order to generate a selective advantage, and also how one key mutation can open up a whole new landscape of evolutionary possibilities. Indeed there are many examples in evolution in which beneficial alleles can rapidly sweep through a population, producing sudden bursts of evolution, sometimes associated with speciation.[9]

The Red Queen

The Red Queen Hypothesis has proved to be a fruitful idea when considering the tempo of evolution. The term derives from Lewis Carroll's *Through the Looking-Glass* in which the Red Queen remarked, "It takes all the running you can do, to keep in the same place." Applied to evolution, the idea is that organisms in active competition with other organisms have to keep evolving just to keep pace with their environment. This is strikingly apparent in the "arms race" of host-parasite relationships in which the host is evolving to protect itself against the parasite and the parasite is seeking to outwit the host.

Once again, bacteria provide us with a vivid example of how this happens in practice. Parasitic viruses that infect bacteria are called Phage, and scientists have studied Phage Φ2 (pronounced "Phage Phi 2")[10] and the bacteria that they infect in order to better understand their evolutionary arms race. The rate of evolution as measured by genetic variation was far higher in the Phage when it was cocultured with the bacteria continually ("coevolution"), rather than when different batches of the same genetically identical bacteria were independently infected with the Phage for the same time period. Furthermore, four Phage genes in particular evolved much faster than other genes (by mutating), showing that they were actively being selected. Perhaps not surprisingly, these were all genes involved in host infection, for the Phage can only replicate by first infecting the bacteria. So coevolution seems to stir things up, preventing organisms from getting stuck in a rut and driving evolutionary change.

But the bacteria still stayed bacteria in this experiment, just as they did in Lenski's culture flasks over decades of evolution. To see how genetics is involved in the transition of one species into another, we have to stand back a little and look at the evolutionary tree of life as a whole.

CHAPTER 6

Genetics and the Evolutionary Tree of Life

DARWIN FIRST DREW a sketch of evolution as a very rudimentary tree of life in one of his early notebooks (1837), and it then became a more sophisticated figure in *On the Origin of Species*. In fact, it was the *only* figure in a book that was destined to change the face of biology, so flashy illustrations have not always been essential for a book to make a big impact. Today we might wish to choose a "bush of life" as an even more appropriate image for evolutionary history (Figure 6.1). It fits well with the phrase that Darwin used to describe what we now call evolution: "descent with modification."

If we imagine all the species that ever lived as comprising branches and twigs on one enormous bush, with mammals and flowering plants on the branches nearer the top, and single-celled organisms like bacteria and yeast nearer the bottom, then we immediately notice three things. The first is that the species living today are not evenly representative of all the species that ever lived, for the simple reason that many of them have gone extinct. But despite that, there are still enough living species to take us back quite close to the origin of life, enabling us to sequence their genomes and construct a "bush" based on genomics.[1] The second point we notice is that the roots of the bush represent living things that must have existed during the emergence of life and the first cells.

Third, we note that the genomes of all these living organisms are connected. If the bush model is correct, then we should see many genes from simple unicellular organisms near its roots being found

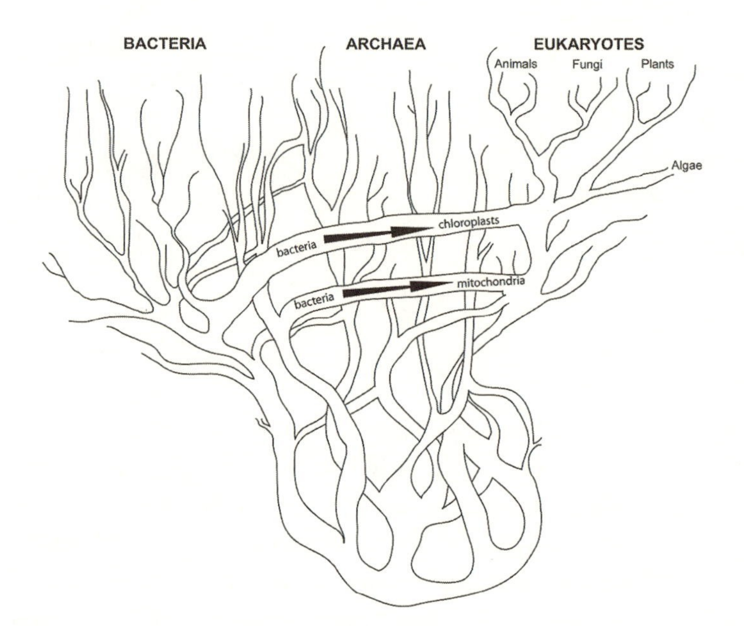

FIGURE 6.1. The image of the bush of life illustrates the interrelationships among the three domains of living organisms (Bacteria, Archaea, and Eukarya). Note the tangled roots of the bush of life, which indicate prevalent lateral gene transfer during the early evolution of living organisms. Note also the whole genome transfers of bacteria that became the organelles (mitochondria and chloroplasts) of the eukaryotes.

in the mammalian and plant twigs at the top, albeit darned like socks because of their great age. That is exactly what we do see. At the same time, we should see new genes appearing at the top that we don't see at the bottom, which is also observed. Besides all that, if stretches of retroviral and other parasitic, nonfunctional types of DNA start appearing at different heights of the bush, then we should see these "flowing upward" to the higher branches, but not appearing in the lower branches, and that is indeed the case. In fact, that latter point is so important that we are going to consider it separately when we come to consider the question of our own evolutionary history.

The bush-of-life analogy has been somewhat challenged by the ubiquitous nature of lateral gene transfer in the eukaryotes and other unicellular organisms. The three great domains of life on earth are the Bacteria, the Archaea, and the Eukarya. Most of these consist of single-celled organisms, in which lateral gene transfer has been rife. Multicellular organisms, in which such gene transfer appears to be less prevalent, only started flourishing during the last 600 million years, less than one-fifth of life's evolutionary history. A bush with branches and twigs implies vertical transfer of genes, but this is clearly not the case with bacteria and with many unicellular eukaryotes, which enjoy both vertical and lateral transmission of genes. It might therefore be best to think of the bush as referring mainly to organisms like plants and animals that we can actually see without the use of a microscope, recognizing that this bush sits on top of a huge tangle of intertwined roots, stretching all the way down to one single common root of life—intertwined roots representing the evolutionary history of life on earth for around 3 billion out of its total 3.8 billion years.

On the bush of life, things are certainly a lot bigger on its upper branches than they are further down. Living things on earth were really small for a very long time: being big is a recent phenomenon. Biologists would get very excited if they could travel back in a time machine to our evolutionary past, but for the ordinary observer it might look extremely boring, because during the first 2.5 billion years of life on earth (approximately), things rarely grew bigger than 1 mm across, about the size of a pinhead.[2] No birds, no flowers, no animals wandering around, no fish in the sea, but at the genetic level lots going on, with the generation of most of the genes that were later used to such effect to build the bigger, more interesting (to us) living things that we see all around us today.

Not until the advent of multicellular life did living organisms start to get bigger, although even then they were generally on a scale of millimeters rather than centimeters. With the flourishing of the late Ediacaran fauna (named after the Australian hills where their fossils were first found) during the period from 575 million to 543 million years ago, we finally move into the centimeter scale. Only in the so-called Cambrian explosion during the period from 525 million to 505 million years ago did sponges and algae grow up to 5 to 10 cm across, and the size of animals began to increase dramatically from that time onward. The Cambrian explosion has drawn particular attention because virtually all the animals with which we are familiar today are derived from that "explosion" of new animal life forms and body plans.

The Cambrian explosion is an example of adaptive radiation, in which a new life form appears and evolves to fill multiple ecological niches relatively rapidly. Organisms fill a niche by making a living in a unique way, whether that means exploiting a new food source or a new strategy for avoiding predators. For example, the emergence of insects during the Devonian and Carboniferous periods (during the period from about 400 million to 350 million years ago) was followed by a huge proliferation in insect types. The radiation of the flowering plants that took place in the late Cretaceous period (after 144 million years ago), and the radiation of mammals in the early Tertiary period (after 60 million years ago), both followed a similar pattern. The generation of novelty leads to adaptive radiation because novelty opens the way to new opportunities for organisms to flourish in particular ecological niches, just as the bacteria in Lenski's experiment described in the previous chapter gained a whole new lease of life once they discovered how to feed on citrate.

The Genetic Code and the Bush of Life

How can we be so sure that all of life's diversity is historically connected? One of the most powerful pieces of evidence for the bush of life, including its tangled roots, is the near universality of the genetic code in all living organisms. Whether archaea, bacteria, or eukaryotes, all living things use the same set of sixty-four DNA codons (introduced in chapter 2) to specify amino acids, pointing to a DNA-containing ancestor right at the deepest root of the bush of life. In some organisms a few of the sixty-four codons specify different amino acids from normal, but the exceptions are rare enough to draw attention in scientific journals and generally prove the rule by shedding further light on evolutionary mechanisms.

Many of the unusual variant codons have been found in mitochondrial rather than nuclear DNA, and the differences are sometimes shared between the mitochondria of vertebrates and evolutionarily ancient organisms such as amphioxus (lancelets). The latter shared a common ancestor with the vertebrates more than 520 million years ago, pointing to some selective advantage; otherwise it is unlikely that the alternative usage would have been maintained.[3]

The evolutionary consequences of an alternative codon have been directly investigated in certain *Candida* yeast species that decode the standard leucine CUG codon as serine, thereby reprogramming the identity of approximately thirty thousand CUG codons that existed in the ancestor of these yeasts and having a profound effect on their evolution.[4] This can occur if a tRNA that binds CUG with its anticodon evolves to bind to serine with its acceptor arm (see chapter 2 for details on tRNAs). By genetic engineering of the yeast, this interpretation of the CUG codon was switched back to the normal leucine, which not surprisingly resulted in a viable but much modified organism, providing strong evidence of codon switch as a mechanism for speeding up yeast evolution.[5]

Our Ancient Genes

One of the fascinations in the bush of life is to compare our own genomic sequence with that of organisms that lie at much lower levels in the bush. In doing so, we realize that much of the informational groundwork for building the cells that constitute our own bodies was carried out millions or even billions of years ago. All cells need to do certain housekeeping jobs like divide, repair damaged DNA, break down food and use it to supply energy, grow, defend themselves from outside attack, and so forth. Once genes evolved to do this, many of them stayed recognizably the same, or at least very similar, throughout billions of years of evolutionary history. If genes are doing a good job, why change them?

We have many examples of such genes. For example, the coral *Acropora* (the material called coral is made by these small anemone-like animals) flourished in the Precambrian before 540 million years ago. A sampling of thirteen hundred gene sequences from this ancient life form reveals that no less than 90 percent are present in the human genome, suggesting that many genes thought to be specific to vertebrates in fact have much older origins.[6]

Another example comes from yeast. For a long time, fungi were thought to belong to the plant kingdom, but genetic analysis has shown that fungi make up a kingdom of their own that is more closely related to animals than plants. Yeast are single-celled fungi, and as eukaryotes they divide in a very similar way to cells found in multicellular organisms like us. Use of this basic insight won geneticist Paul Nurse a Nobel Prize. Cancers are caused by cells that can't stop dividing. Nurse took the risky career step of deciding to investigate how normal cells divide using yeast as his model system. Many people at the time (1980s) thought that yeast cells would turn out to be very different from animal cells and would tell us little about what goes wrong in cancer. In fact, exactly the opposite is the case. Nurse, along with others, discovered how a key class of enzymes called kinases regulate every stage of cell division, a vital insight for those trying to understand cancer. It now turns out that this basic tool kit of gene-encoding proteins is found in virtually every nucleated cell in the world, including our own, regulating how cells divide. Nurse won a Nobel Prize, became president of the Rockefeller Institute in New York, and then president of the Royal Society, Britain's premier scientific society. Choosing the right questions to ask is often the scientist's biggest challenge.

So, as these examples illustrate, a different kind of gene flow is going on here from the one mentioned in the previous chapter—

the flow of ancestral genes up the trunk and along the branches and twigs in the one great history of life. And all of the many mechanisms for generating genetic diversity reviewed in chapter 4 have been involved in constructing the bush.

Gene Duplication and the Bush of Life

One key mechanism that has played a vital role in constructing the bush is gene duplication. The spare gene so generated is less likely to be under the pressure of natural selection because its parental gene is already in the genome doing a good job. The newly duplicated gene may thus start to accumulate mutations and acquire a new function, which is then subject to natural selection.

An analogous situation can occur during chromosome doubling in which whole genomes duplicate themselves. This activity generates huge possibilities for further novelty, because now not just one duplicated gene but a whole genome is free to "wander off" mutationally and do new and exciting things. In practice the big genome is eventually trimmed down to a more manageable size, but can now be quite different from what it was before. Although whole genome duplication happens only very rarely, such dramatic events have played an important role in evolutionary history. For example, the entire yeast genome is thought to have duplicated about 100 million years ago.[7] Wheat is hexaploid, meaning that it has no fewer than six copies of its entire genome. Backing up your computer data is always a good idea, although six backups seem somewhat excessive! Of more personal significance to us is that vertebrate evolution seems to have involved two complete genomic duplication events that clearly have huge significance for our own evolutionary history.[8]

Thinking about alcohol digestion can help us see how the duplication of even a single gene can contribute to human welfare (or not). Those who consume alcohol need an enzyme called alcohol dehydrogenase (ADH) to convert the alcohol into acetaldehyde, but the original gene encoding ADH seems to have evolved in yeast to catalyze precisely the reverse reaction. Following duplication of this ancestral gene, one of the two duplicated genes was then free to mutate slightly until it became good at using alcohol as its starting material. In fact, the ancestral gene has been made artificially by reverse engineering the ADH gene.[9] All of our own five different types of ADH appear to have evolved by repeated gene duplication events starting from the single original ancestral gene. So we should be thankful to yeast twice over: first for giving us some great beers, as of course it is yeast that ferments sugar into alcohol, and also for bestowing upon our own evolutionary lineage the ADH that we need to metabolize the alcohol once we drink it.

The development of scales in certain kinds of fish is also a story about gene duplication, this time turning into a story about food rather than drink. Zebrafish, a popular freshwater aquarium fish with striking zebralike blue stripes, are, like *Drosophila,* a favorite of geneticists due to the ease with which they can be genetically manipulated. Walking into many geneticists' laboratories today is like walking into an aquarium shop, with tanks of zebrafish stacked up against the wall. As many as 20 percent of genes that are present as single copies in mammals are found as duplicates in zebrafish. One of these genes, called *fgfr1,*[10] is really important for development both in fish and mammals. Without this gene mice do not develop at all, but zebrafish, surprisingly, can develop normally even when it is mutated. The reason is that zebrafish have a second backup gene that compensates for a lack of *fgfr1* during embryonic development. The duplicate gene has also evolved a more specialized function of its own: to regulate how many scales a fish has in adult life. When *fgfr1* is kept normal and the duplicate copy is mutated, the fish develop normally, except that they almost completely lack scales.

This second copy of *fgfr1* likewise regulates scaling in a close relative of zebrafish, the domesticated common carp. In medieval times, carp were widely used as a substitute for meat during Lent. Over hundreds of years, strains of carp with far fewer scales than the original were selected, making them easier to prepare for cooking. Unknown to them, the monks breeding carp with fewer scales in the garden ponds of their medieval monasteries were actually engaged in primitive genetic engineering: they were breeding fish carrying mutations in the duplicated copy of *fgfr1*. The two copies of *fgfr1* provide a good example of the way a duplicated gene can retain its original function and evolve to acquire other functions. In fact, the teleost fish, of which zebra and carp are exemplars, represent about 95 percent of all living species of fish and display a huge range of diversity in structure and physiology. Gene duplication has very likely played an important role in generating this diversity.

Gene duplication represents an important driver of evolutionary change. New information is constantly pouring into the bush of life as duplicated genes mutate to acquire new functions. Gene duplication is one of the key "fertilizers" that help the bush to flourish.

The Genetics of Speciation

A species refers to a population of organisms that interbreed with each other but generally not with other organisms, and are therefore said to be "reproductively isolated." Clearly this definition is only useful for living species. Extinct species have to be defined based on the fossil record by looking at differences in morphology (body shape, plan, size, etc.), so classification of dead species can be less reliable.

Speciation is thought to occur either by allopatric mechanisms, which geographically divide a population (such as the separation of two landmasses as a result of continental drift), or by sympatric mechanisms, which reproductively isolate two subpopulations that inhabit the same geographic area.

Phylogenetic trees are diagrams that show how species are related to each other. Based both on morphological and DNA sequence data, they help us visualize the bush of life. A typical phylogenetic tree is illustrated in Figure 6.2. The arrowed numbers refer to millions of years before the present and indicate the estimated dates at which last common ancestors are predicted to have lived. The last common ancestor of any set of species is symbolized by the node at which all of the different lineages in the set converge, as if it were possible to run them backward in time. We (*Homo sapiens*) are second down in the list of species on the right, and our last common ancestor with the mouse (*Mus musculus*, fifth down) is estimated at 75 million years ago; with the horse, 92 million years ago (*Equus caballus*, ninth down); and with the *Drosophila* fruit fly, 700 million years ago (*Drosophila melanogaster*, thirteenth from the bottom of the list). The further back in time we go, the greater the approximation in estimating the dates.

In general, estimates of time based on phylogenetic trees of this kind fit quite well with those constructed based on fossil data, but this is not always the case. Variant estimates are not unexpected, since fossils tell us when a particular species lived and went extinct, supported by a range of solid dating methods. Phylogenetic trees, on the other hand, predict last common ancestors, providing dates that might well be much earlier than the time at which a particular species flourished in sufficient abundance to leave its fossils behind for posterity. Generally only a tiny percentage of animals and plants become fossilized.

For speciation to occur, a genetic barrier of some kind must come between different organisms so that they can no longer exchange their genes through sexual reproduction: gene flow can no longer occur. Given the huge number of different types of genetic variation, finding an equally large number of ways in which speciation takes place at the genetic level is not surprising. For example, the chromosomes might no longer pair up properly during fertilization; speciation might involve a mutation in a gene encoding a sex pheromone—wrong smell, no sex; it could involve anatomical differences such as the width of the birth canal, and so forth. The list is endless.

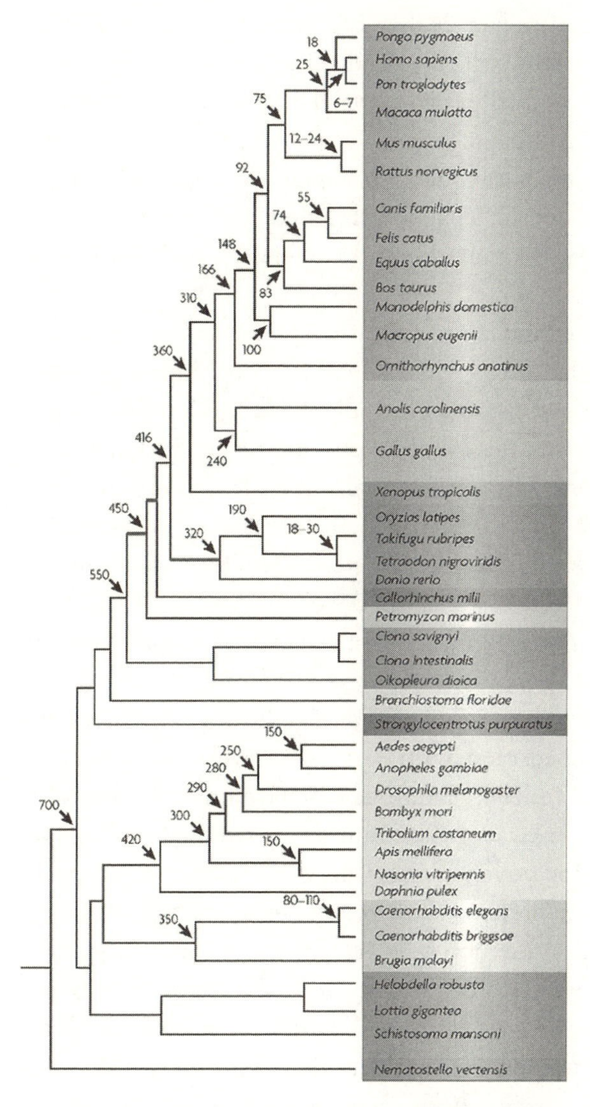

FIGURE 6.2. This evolutionary tree shows relationships among animals based on genetic sequencing information. Scientific names are listed to the right of the tree. Estimated divergence times (million years ago) are indicated at selected nodes. Divergence times and branching order may yet be revised as new findings and data arise. Reprinted by permission from Macmillan Publishers Ltd. Ponting, C. *Nature Reviews Genetics* 9 (2008): 689–98.

The initial genetic changes that establish a breeding barrier between two new species will only be a small percentage of the genetic differences that eventually evolve between the two species with the passage of time. Although genomic sequencing allows us to compare the complete DNA sequence between two closely related species, that in itself doesn't tell us which genetic differences represent the key steps that led them to go their separate ways in the first place.

In some plants and a few animals, sympatric speciation can occur suddenly by the addition of extra sets of chromosomes to the genome, an event known as "polyploidy," which can happen in the following way. Occasionally one species of plant can fertilize another to form a hybrid. Normally hybrids are sterile, just as a female domestic horse and a male domestic ass can mate to generate a mule (a hybrid), which is also sterile. However, unlike mules, hybridization in certain plants can be followed by doubling of the chromosome number, giving rise to a new species due to reproductive isolation.

Many flowering plant species are thought to have arisen by polyploidy, so rapid speciation of this kind is relatively common in the plant kingdom.[11] Indeed, plant breeders have utilized polyploidy for years to create new and often commercially valuable plant species artificially. The pollen of one plant is painted on the stigma of another from a different species, and the resulting hybrid plant is then treated with a chemical called colchicine to cause polyploidy, leading to the formation of a new species. The first artificially created hybrid species was a primrose called *Primula kewensis* that has thirty-six chromosomes and was derived from two different parental *Primula* species, both having eighteen chromosomes. In fact, in this case colchicine wasn't used as the chromosomal doubling happened spontaneously.[12] To see the fruit of recent speciation events, we need only look out of the window, for about a third or more of the plants in our gardens have likely been created by artificial polyploidy.

Polyploidy is much less common in animals than in plants, but among animals it is most common in species that have parthenogenetic females. Parthenogenesis is the growth and development of an embryo without fertilization by a male. Such species include some shrimps and moths, as well as certain lizards, beetles, fish, and salamanders. Polyploidy has also been found in some fish and amphibians that reproduce sexually.

Identifying specific genes that result in speciation is a more difficult task than identifying structural chromosomal changes that have led to speciation. Nevertheless, there is a growing list of candidate "speciation genes." An interesting example is found among the pretty *Mimulus* flowering species that display "a stunning variety of life history, developmental and physiological traits in response to a broad range of habitats."[13] As is true for many flowers, pollinating insects shape *Mimulus* diversity perhaps more than any other evolutionary influence: insects are the real gardeners, which explains why the "Yellow Upper" (*YUP*) gene region is a good candidate "speciation gene." *YUP* controls the absence of yellow pigment in the pink petals of *Mimulus lewisii,* which is pollinated by bumblebees, and the presence of yellow pigment in its red-flowered sister species *Mimulus cardinalis,* which is pollinated by hummingbirds.[14] If the *YUP* gene regions of these two species are swapped, they switch colors and pollinators; the insects follow familiar colors to different flowers. In such ways can living organisms regulate gene flow by means of a single variant allele.

The cichlid fishes of the African lakes provide a well-studied example of rapid speciation in animals, a classic case of "adaptive radiation"; more than a thousand different species have evolved in these lakes over the past 1 million years. In Lake Victoria alone, around five hundred species have evolved in the past fifteen thousand years, about one new species every thirty years on average.[15] Amazingly, the lakes' cichlids account for more than 10 percent of the world's freshwater fish species, and the cichlid species in one lake are quite distinct from those in another.[16] Different species show multiple differences in morphology linked to their feeding habits. Virtually every major food source in the lake is exploited by one species or another: some cichlids eat insects, others crustaceans, others eat plants, and yet others mollusks. Each new species has found its particular ecological niche. One species, *Haplochromis welcommei*, has the odd habit of feeding on fish scales, which it scrapes off the tails of other fish!

The genetic variation underlying these recent, multiple cichlid speciation events is gradually being elucidated.[17] For example, the gene that encodes "long wavelength-sensitive opsin," a protein important in color vision, comes in different variants that are thought to be important for speciation via female mate choice. As with flowers, if you're a fish with the "wrong" color, then your days with your "old mates" as a member of your original species may be numbered. Other candidate speciation genes encode the pigments that provide a wonderful range of different cichlid skin colors, as well as the genes that determine jaw morphologies and tooth shape. Barriers to gene flow may arise from many different mechanisms.

One key genetic mechanism in vertebrates involves the "speciation gene" *prdm9* which regulates whether or not hybrids from matings between genetically closely related animals are sterile or not.[18] If they are sterile, then this means that genetically distinct but related animals have become "reproductively isolated," a first key step towards speciation. The *prdm9* gene encodes a DNA binding protein that recognizes specific motifs within the DNA that undergo very rapid evolution. A mismatch between the protein sequence and its recognition site in the DNA means that the chromosomes can no longer pair up properly, so there are no viable offspring. Once the reproductive barrier is in place, then the way is opened up for the development of two distinct species.

Like a musical score, the tempo and pace of speciation seem to vary considerably, and the cichlids are definitely a case of *fortissimo*, whereas some snail species are, well, just snail-like in their rate of speciation, remaining pretty much the same (in the fossil record) for millions of years until a sudden burst of speciation occurs. Of course, "sudden" in the geological record may refer to tens of thousands of years or more. This uneven record gave rise to Eldredge and Gould's famous theory of "punctuated equilibrium," which suggests that species go through long periods of stasis, followed by rapid bursts of novelty.[19] Novelty may be triggered by changes in the environment; for example, it may suddenly become wetter, drier, warmer, or colder in a particular area, and new selection pressures then come to bear on all the affected organisms.

Evolution involves a finely tuned, lawlike balance between chance and necessity in which genomes play the role of the musical keyboard in generating biodiversity. But whether the pace of the music is fast or slow, genetic variation plus natural selection together ensure that each ecological niche is used for the optimal benefit of the organism.

Speciation and the Genetics of Evo-Devo

The key role that developmental genes play in body building has already been introduced in chapter 3, illustrated by the Hox "master genes" that regulate body segmentation. The discovery of genes like these kick-started the modern study of the evolutionary significance of genes that regulate embryonic development, called evo-devo for short.

Clearly not all genes are equally open to change. If a single copy of a gene encodes a protein that is essential for life, and it is operating already with great efficiency in a particular organism, then all that will happen if it mutates will be a dead organism, which is not very helpful. Such genes are sometimes known as "housekeeping genes": the "house" simply won't exist without them. Such genes are under very strong selection and stay virtually the same across billions of years of evolutionary history.

Take cytochrome c, for example, a protein around one hundred amino acids long found in mitochondria, important for the energy-generating capacities of these vital organelles. Knock out a key component of your local power plant, and your city will be plunged into darkness. Mess with the sequence of cytochrome c, and energy generation will likewise close down; in other words, the mutation will be lethal. This is why the cytochrome c found in animals, plants, and many unicellular organisms is remarkably the same. At the amino acid level, our cytochrome c and that of chimpanzees is identical. It is likewise identical between pigs, cows, and sheep. Even a penguin and a tomato display only eighteen amino acid differences.

So in practice some genes are more "evolvable" and others are much less so. It is not that housekeeping genes do not evolve. Rather, once they have evolved to be so critical for life, they change very slowly and cautiously. Genes that are far more likely to play key roles in evolutionary change act as input/output genes or hotspot genes, encoding key switching proteins that integrate whole sets of information that are then mediated to downstream effectors. These proteins are often transcription factors (introduced in chapter 2), which integrate various signals and orchestrate the actions of many other genes. For example, a gene that delights in the name of *shavenbaby* regulates the existence and distribution of fine trichomes or cellular hairs on the surface of the larvae of *Drosophila*.[20] Mutations in *shavenbaby* can lead to a lack of hairs—hence the name.

Hotspot genes are evolutionarily fertile because they run genetic modules, or miniprograms, which can be variously deployed, like the *shavenbaby* program for converting nonspecialized cells into hair-making cells. The mutations that result in "shaven babies" occur in the gene's regulatory sequences that determine how much of the protein is made. Many evo-devo studies have likewise shown that the regulatory regions of genes, together with the great array of transcription factors involved in on-off switching, provide the principal engine for evolutionary change. Studies on sticklebacks, and on the Hox genes in different species, provide some useful examples of how this works in practice.

Evo-Devo in Sticklebacks

Threespine stickleback fish, relatives of seahorses, are like cichlids in that they have evolved into many distinct forms in different lakes. The original threespine stickleback lives in coastal seas and, like salmon, swims up rivers to spawn. During the last big retreat of the glaciers ten thousand to twenty thousand years ago, many populations of threespine sticklebacks were trapped in newly formed lakes, and they have since adapted to those environments.[21] Whereas marine sticklebacks are heavily armored with an array of plates and spines (the species is called *Gasterosteus aculeatus,* meaning "bony stomach with spines"), the freshwater populations all show versions of armor loss. The pelvic spines in particular seem to be lost in environments with predators like dragonfly nymphs that can catch young fish by the spines.

How could genetic variation evolve in such a short time to generate such different phenotypes without causing changes lethal to the sticklebacks along the way? Investigations into the genetic basis of pelvic spine loss have turned up a gene called *Pitx1,* a transcription factor that regulates the developmental program leading to spine formation. The same gene is expressed in the hindlimbs but not the forelimbs of mouse embryos, and indeed in its absence mice are born with dramatically reduced hindlimbs. Given that the freshwater stickleback *Pitx1* gene is required for many aspects of development, how does its mutation regulate stickleback armor so selectively? It turns out that a DNA lesion in freshwater sticklebacks has removed about five hundred base-pairs of regulatory sequence that normally upregulates Pitx1 protein production specifically in the pelvis. With the Pitx1 protein expressed normally in all other tissues, but virtually lost in the pelvic region, development proceeds normally in all other respects.

Most remarkably of all, mutations in the same or closely related

Pitx1 gene regulatory regions have evolved independently multiple times in stickleback populations found in different freshwater lakes around the world. These populations exist in landlocked lakes where there can have been little chance of swapping fish between the lakes. How could the very same or highly similar random genetic changes evolve so rapidly in different fish populations? The answer lies in the fact, discussed in chapter 4, that mutations are not really random: there are regions of the genome where variation is much more likely to occur. The gene *Pitx1* is located in a fragile region containing many repeats that is subject to more deletions than other regions.[22]

So these fish are predisposed to remarkably rapid changes in their pelvic spines at multiple levels. The relevant regulatory gene is located in a highly changeable area of the genome; the gene in question can act as a master-control body armor–switching apparatus; and variants that reduce pelvic spines will be rapidly selected in environments where those spines decrease fitness.

What makes the story even more interesting is that another genus of sticklebacks, the ninespine sticklebacks, has undergone the same type of isolation in lakes and loss of pelvic spines as the threespine sticklebacks. However, the gene that is responsible for pelvic reduction in ninespine sticklebacks is not *Pitx1*, but rather a different gene on a different chromosome. Loss of pelvic spines in these two genera of sticklebacks by different mechanisms is an example of convergence in evolution, a topic we return to at the end of this chapter.

Evo-Devo and Hox Genes

No discussion of evo-devo would be complete without at least some further mention of our old friends the Hox genes, which also illustrate well the notion of "deep homology."[23] Like the notion of "deep time," a phrase often used when looking back into the earliest history of our universe, deep homology becomes apparent when we peer back into our evolutionary past and find that the same genes and regulatory networks keep cropping up in animal and plant bodies. Ancient regulatory systems both constrain organisms and provide a basis from which novelty can spring.

The Hox genes were right there in the earliest bilaterians at least 600 million years ago. Bilaterians include all animals that have a recognizable front end and back end, as well as an upside and a downside, which describes 99 percent of described animal species. Sponges and jellyfish don't count; they have a topside and a bottomside, but no front or back. If you see a jellyfish floating toward you while swimming, its radial symmetry can be a little unnerving ("why can't it be clearer which way it's going?"). This is why the best sci-fi movies always have nonbilaterians, squidgy things, which look thoroughly unlike us, rather than cute creatures like ET, or skinny avatars, which are certainly bilaterian (and also make out under trees).

The first bilaterians were discovered in rocks dating from 580 million to 600 million years ago from Guizhou Province in China.[24] There tiny fossil bilaterians were identified (named *Vernanimalcula*, meaning "small spring animal"), only the width of a few hairs across, but with the classic mirror-image body structure of a bilaterian. These little creatures probably zoomed around on the sea floor feeding on microbes which at that time covered the floor in great mats.

The Hox genes present in all the bilaterians studied so far are intimately involved in the development of their body plans, just as we saw in the case of the segmentation of *Drosophila*. They are typically grouped in clusters, and the sequence of the Hox genes in the genome parallels the front-to-back structure of most bilaterians, again just like *Drosophila*. The many Hox genes that control an array of body-building programs in different animals represent a classic case of gene duplication followed by diversification. Their numbers vary hugely. The Japanese eel has at least thirty-four distinct Hox genes, the large number arising from duplication of the whole genome, a privilege shared by zebrafish and pufferfish.[25]

Further diversification of Hox genes is sometimes helped along by unusually high numbers of transposons being inserted in the Hox gene region, forcing a reorganization of the gene regulation system, with important implications for the evolution of snakes and lizards.[26]

But what happens when we really go back deep into evolutionary time? There are no Hox genes in bacteria or yeast, nor in plants, but they are present in all of the nonbilaterian animals—the sponges and Cnidarians—which are represented by more than nine thousand species of aquatic animals, including jellyfish, sea anemones, and corals. Cnidarian fossils have been found dating from about 580 million years ago, so they predate the Cambrian explosion when vertebrate body plans began to be distinguished from the invertebrate. The Hox genes of the nonbilaterians, however, are scattered around the genome rather than organized into clusters as they are in bilaterians.[27] Exactly what the Hox genes in sponges and Cnidarians do remains to be figured out, but they don't appear to control body plans. So it seems as if the secret to being an animal, at least an animal with a developed intelligence, is to assemble a tool kit of Hox genes from precursors that originally had other functions, organize them, and then modify them to produce one of the myriad variants of animal body plans that we see around us on our planet today. Understanding the modification of genetic tool kits is what evo-devo is all about.

Genes, Convergence, and Constraint

As already highlighted, evolution arises from the balanced interplay between chance and necessity. Chance generates the raw material of genetic variation. Necessity is represented by the winnowing filter of natural selection. Too much chance, and stable species with distinct characteristics would never evolve. Too much necessity, and biological systems would remain static, frozen in time. Chance and necessity play an ongoing duet in the music of life.

But in the end, necessity has the upper hand. The reason for this lies simply in the constraints of living a carbon-based life on a planet with the particular properties of gravity, wet and dry, hot and cold, high and low, which generate a fixed range of ecological niches. Evolution is like a search engine exploring design space, and only a certain number of solutions are available to living a carbon-based life in a given ecological niche, solutions that evolution keeps discovering again and again.

This process leads to the remarkable phenomenon of convergence, whereby the same biological adaptations emerge repeatedly in independent evolutionary lineages. At the phenotypic level these can be very striking.[28] The hedgehog tenrecs of Madagascar were long thought to be close relatives of "true" hedgehogs, because their respective morphologies are so similar, but it's now realized that they belong to two quite separate evolutionary lineages and have converged independently upon the same adaptive solutions, complete with spikes. Compound and camera eyes taken together have evolved more than twenty different times during the course of evolution. If you live in a planet of light and darkness, then you need eyes—so that's what you're going to get! There are hundreds of other examples: evolutionary convergence is the rule rather than the exception.

Evolutionary convergence at the phenotypic level does not mean that a complete set of new genes evolves separately each time to build, for example, an eye. Instead genomes are like garages or basements where, instead of putting the car, we store (ok, males store, let's be honest here) all kinds of bits and pieces, "just in case they might come in useful one day"—as sure enough they do. So genomes contain genes that may be switched off, ready for use at some future time as required, or genes that presently have quite different functions which can be pressed into service. Rummage around in the genomic garage and you can find surprising things. There are many examples of genes that encode "moonlighting proteins"—proteins that carry out quite different tasks depending on whether they are inside the cell or outside, the particular tissue in which they are located, or even which specific location they occupy inside a cell.[29]

The gene that encodes the transcription factor Pax-6 provides a dramatic example of a gene that has been involved in photosensitivity and eye building for more than 500 million years. Pax-6 was first identified in humans because it causes the disease Aniridia, underdevelopment of the retina leading to blindness. When the Pax-6 gene is disrupted in mice or *Drosophila,* it likewise causes blindness, and when the mouse Pax-6 is incorporated by genetic engineering into blind *Drosophila,* then sight is restored. This is rather remarkable because *Drosophila* have compound eyes, consisting of multiple single photoreceptor units, and mice have camera eyes, a quite different kind of structure with a single refracting lens in each eye. Yet the Pax-6 genetic control lever, when "pulled," can switch on a whole sequence of eye-building molecular events during development of both organisms.

It turns out that Pax-6 is expressed in eyes right across the animal kingdom, in vertebrates, arthropods, annelids, and mollusks. Other Pax family genes are present in the photoreceptor cells, used to detect light, of the Cnidarians. The box jellyfish, which has camera-type eyes with a cornea, a lens, and a retina—just like the eyes found in us and in other vertebrates—uses a kit of eye-building genes, including a family member called Pax B.

Other genes illustrate the way in which convergence to generate similar adaptations operates at the molecular level. Echolocation is the method that mammals such as bats, porpoises, and dolphins use for hearing. It involves sending out pings of sound that bounce off objects and are then received back and analyzed—an animal sonar system used to locate and identify prey. The brain works out how long it takes for the ping of sound to come back, and how far away the object is. If you've ever watched bats wheeling around the trees at dusk, or perhaps around your head in a dark cave, then you'll know that the system has to work incredibly quickly and efficiently to avoid mishaps. Bats can detect the presence of a tiny crawling insect or even a human hair, and can recognize each others' voices.

A special protein called prestin is key to this sophisticated high-speed process. The gene that encodes this protein is unique to mammals and has evolved independently several times since mammals split off from the birds in evolutionary history more than 100 million years ago. Prestin is found in the outer hair cells of the inner ear of the mammalian cochlea, a fluid-filled chamber. As the sound waves the ear receives compress the fluids, so the sensory hairs surrounding the chamber move very slightly and convert their movements into nerve impulses via thousands of hair cells. The outer hair cells that serve as an amplifier in the inner ear refine the sensitivity and frequency selectivity of the mechanical vibrations of the cochlea.

The specialized prestin found in echolocating mammals provides a much faster system for converting air pressure waves into nerve impulses than the prestin found in mammals (like us) that do not use echolocation. The story of convergence became apparent when it was discovered that the prestin gene has accumulated many of the same mutational changes in bats, porpoises, and dolphins, changes that are essential for the prestin protein to perform its unique functions.[30] Similar changes have occurred in unrelated lineages of different bats. Genetic evidence suggests that these changes have undergone natural selection. In other words, here is an adaptation of great advantage to the animal that has it, so animals carrying this particular set of mutations in the prestin gene are more likely to reproduce and spread the beneficial gene around an interbreeding population. The particular advantage may well be the necessity to hear very high frequencies, far above the ability of the human ear. The advantage of possessing this fancy piece of echolocation equipment has helped shape the evolution of the prestin gene such that it has converged on the same adaptive solution independently on multiple occasions.

But how can a gene navigate its way to generate a protein that has just the right function for the job? What about all the intermediate steps along the way? Is the pathway completely random, so if we played the tape of life again, the outcome might be quite different? The incremental steps are actually highly constrained: only certain paths up the mountain are possible. This has been documented in the evolutionary pathways of certain bacterial genes that can be studied in detail in the laboratory due to their rapid dividing time and the ease of detecting deleterious mutations.

For example, a research group from Harvard published a paper titled "Darwinian Evolution Can Follow Only Very Few Mutational Paths to Fitter Proteins."[31] They studied an enzyme called β-lactamase which breaks down antibiotics such as penicillin. Provided that bacteria have versions of this enzyme that are functioning efficiently, they grow quite happily in media containing antibiotic. If a gene is under natural selection, then it needs to evolve in small incremental steps, each step increasing the fitness of the organism or at least not decreasing it. Five amino acids are needed at five key positions in the sequence of amino acids that make up the β-lactamase enzyme that enable it to function well enough to enable the bacteria to grow in antibiotic. By random events you could imagine the gene evolving to this state through five mutations that might occur in any order; in principle there could be 5x4x3x2 = 120 different mutational pathways to achieve the goal of optimal enzyme efficiency. But in practice the Harvard researchers found that 102 of these pathways are barred because they decrease the fitness of the bacteria, meaning their ability to flourish in the presence of antibiotic; of the remaining 18 trajectories, only a very few were really favored. The authors' conclusion is intriguing: "We conclude that much protein evolution will be similarly constrained. This implies that the protein tape of life may be largely reproducible and even predictable."

It seems that there is only a limited number of ways in which carbon-based life can exist on a planet with these particular properties, and the evolutionary search engine will keep finding these solutions for living again and again. The idea of "fitness landscapes" can be quite useful for envisaging how this happens at the molecular level, as Figure 6.3 illustrates. Only a few ways along the slopes and valleys arrive at a particular protein function (the "peaks") because only some genetic mutations will get you there and not others. An evolutionary path seems to be mapped out in front of the gene encoding the enzyme, and the genetic dice are thrown until the enzyme structure is generated that optimizes fitness for its particular function. This is no random process, each step along the way being preserved by benefits to the organism that uses the enzyme.

Overall, around 98 percent of all the amino acids in all proteins apparently cannot change because of the striking decrease in fitness of the organism that would result.[32] So the genes that encode these

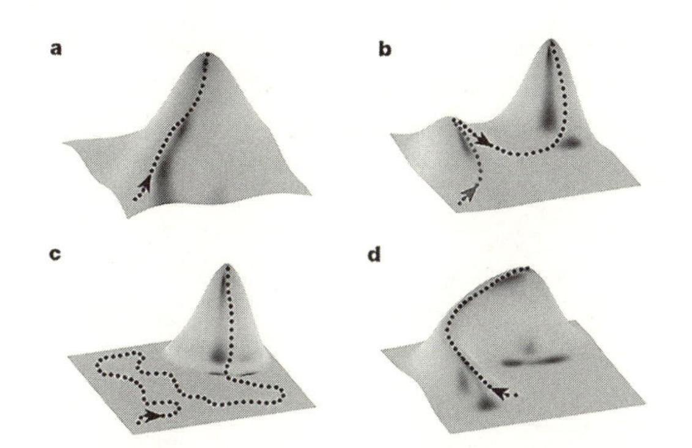

FIGURE 6.3. Molecular fitness landscapes. Fitness is shown as a function of amino acid sequence: the dotted lines are mutational paths to higher fitness. a) Single smooth peak. All direct paths to the top are increasing in fitness. b) Rugged landscape with multiple peaks. The black path has a fitness decrease that drastically lowers its evolutionary probability. Along the gray path, selection leads in the wrong direction to an evolutionary trap. c) Neutral landscape. When neutral mutations are essential, evolutionary probabilities are low. d) Detour landscape. The occurrence of paths where mutations are reverted shows that sequence analysis may fail to show mutations that are essential to the evolutionary history. Reprinted by permission from Macmillan Publishers Ltd. Poelwijk, F., et al. *Nature* 445 (2007): 383–86.

amino acids cannot change either, at least not by mutations that change the amino acid sequence. This might sound like a recipe for a static protein world. In practice, though, proteins do evolve, but they just do so really slowly and cautiously. Indeed, it's unlikely that the evolutionary search engine has yet completed its job of searching the complete repertoire of protein "design space," but it has come a long way in 3.8 billion years, and the present snapshot that we have certainly points to a highly constrained molecular world.

So evolutionary constraint, of which convergence is a well-documented consequence, operates at several different levels in parallel: the phenotypic morphological level, the cellular and protein function level, and at the level of the genes. In reality all these levels are integrated together into one functioning system. Living systems evolve and are constrained in the evolutionary pathways that they can follow.

This discussion brings us to our own little twig on the evolutionary bush of life. The particular constraints that brought about our twig appear to be no different from the constraints that brought about all the others. Yet the objective observer will need to admit that something is very special about our twig.

CHAPTER 7

The Genetics of Human Evolution

How does our own genome compare with those of our immediate ancestors and relatives on the evolutionary bush of life? Sequencing studies on ancient DNA are becoming increasingly sophisticated and have already revealed some of the fascinating genetic similarities as well as differences between us and our nearest relatives.

Our own recent evolutionary history is summarized in Figure 7.1. Although many pieces of the jigsaw are yet to be joined up with certainty, the broad outlines have become well established through the discovery of hominin fossils over the past few decades. "Hominin" refers to those species, now extinct, thought to be on the evolutionary lineage between us and our last common ancestor with the apes. Here we focus only on those hominins that have yielded some genetic data.

There is a strong case that Africa provided the cradle for humanity. *Homo erectus* was the first hominin species to migrate extensively out of Africa, leaving its fossil remains in Asia as far east as Indonesia and in parts of Europe, dating from 1.8 million years to as recently as thirty thousand years ago. About six hundred thousand years ago appeared the *Homo heidelbergensis,* who are thought to have been the immediate precursors of *Homo sapiens,* anatomically modern humans. *Homo heidelbergensis* had an average brain size of around 1,200 cubic centimeters, only 200 cc short of the average size of our own. The fossil bones from *Homo heidelbergensis* are too old to obtain DNA for sequencing studies, but we now have some reliable data on more recent hominins.

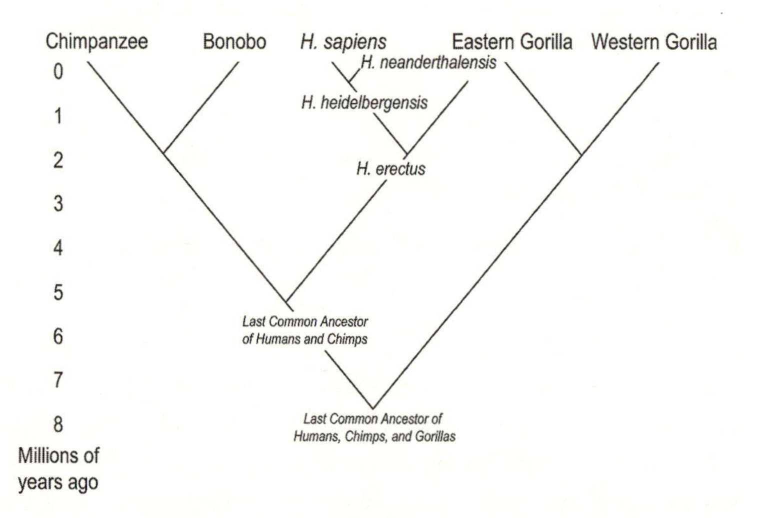

FIGURE 7.1. Humans and their closest relatives. The closest living relatives of humans are the chimpanzees, which consist of two species, the common chimp and the bonobo, and the next closest living relatives are the gorillas. The last common ancestor (LCA) of chimps and humans existed between 5 to 7 million years ago (mya). The lineage linking the LCA of humans and chimps to modern humans is called the hominin lineage, and the fossil record indicates that many species of hominins existed. Three species are shown here: *H. erectus, H. heidelbergensis,* and *H. neanderthalensis,* which is the closest species to *Homo sapiens* yet identified.

About four hundred thousand years ago, a further twig on the evolutionary bush appeared in Europe: the Neanderthals, most likely, like us, descended from *H. heidelbergensis* and were well adapted to the harsh climatic conditions they would have experienced as the northern ice cap advanced and then retreated across Europe in a succession of ice ages. Neanderthals were powerfully built with long skulls, brains as big as humans, huge noses, large ridges over their brows, and barrel-shaped chests, weighing about 30 percent more than modern humans of the same height. Neanderthals are our first cousins, albeit rather scary ones.

Following an extensive European ice age, Neanderthals migrated to the Levant (the area now occupied by the countries to the east of the Mediterranean) from eighty thousand years ago, retreating back into Europe fifty thousand years ago, before finally going extinct about thirty thousand years ago. Their total range was extensive, stretching from the Iberian peninsula (now Spain) in the west to Uzbekistan in the east. Yet their effective population size was probably never very large, around three thousand to twelve thousand individuals based on genetic studies.[1] "Effective population size" is defined as the number of individuals in a population who contribute offspring to the next generation.

With cutting-edge technology and much painstaking work, it has been possible to extract DNA from three Neanderthal bone specimens ranging in age from about thirty-eight thousand to seventy thousand years old and, based on several of these samples, to generate a fairly complete genome sequence.[2] In the same study, five human genomes representing individuals from sub-Saharan Africa, Western Europe, Han Chinese, and Papua New Guinea were sequenced for comparative purposes. A striking result from these comparisons is the very high degree of similarity between human and Neanderthal DNA, much higher than that between humans and chimps. For example, we share our FOXP2 gene with Neanderthals, the gene involved in language discussed in chapter 3. This does not imply that the Neanderthals had language, but had their FOXP2 been the same as the chimps, then it would have made their possession of language less likely. The greater similarity of the human and Neanderthal genomes is to be expected since our last common ancestor with chimps lived about 5 million to 6 million years ago, while our last common ancestor with Neanderthals lived around five hundred thousand years ago.

Features that occur invariably in all humans (which are therefore said to be "fixed" in the human genome), but which are absent or variable in the Neanderthal, are clearly of particular interest in helping to elucidate what makes *Homo sapiens* unique. In all, seventy-eight nucleotide changes that would change the amino acid sequence of particular proteins were found in the Neanderthal DNA that were the same as the chimpanzee sequence but different in the human. Since the Neanderthal genome sequence was incomplete, this is a minimum estimate. The proteins involved range across a wide range of functions, including smell, immune function, cell adhesion, sperm function, skin function, and many others. In addition, significant differences are present in certain microRNAs, together with the Human Accelerated Regions introduced in chapter 2, as well as 111 duplications of small DNA segments that are found in the Neanderthal but not the human sequence—all a good reminder that building a new complex species involves a wide array of different changes.

One of the fascinating contributions that contemporary genetics has made to the hominin evolutionary bush is the uncovering of a potential recent twig even in the absence of any other fossil data. This contribution arose from the discovery by a Russian team of a sliver of finger bone from a remote Siberian cave in the Altai Mountains, known as the Denisova Cave. The team stored it away, thinking it was from one of the Neanderthals that frequented the cave between thirty thousand and forty-eight thousand years ago. But when DNA extracted from the bone was eventually sequenced, the data revealed a population distinct from both humans and Neanderthals.[3]

The finger appears to belong to a novel hominin population that shared a last common ancestor with Neanderthals more recently than humans, and overall is genetically closer to Neanderthals than to humans. The cold Siberian environment had helped to preserve the integrity of the DNA, whereas in the hotter, wetter African climates in which so many hominin specimens have been obtained, DNA degrades much faster. It is too early to say whether the so-called "Denisovans" represent a separate species, and fossil data will be required to clarify that question. Furthermore, the results suggest that Melanesians—the inhabitants of Papua New Guinea and islands northeast of Australia—have inherited as much as one-twentieth of their DNA from the "Denisovans," indicating that some limited interbreeding took place between these ancient populations. The twigs on the hominin evolutionary branches are likely to become even more numerous as genomic studies continue to extend their reach. Most fascinating of all is the idea that multiple hominin lineages were coexisting in Europe and Asia, along with modern humans, as recently as twenty thousand to forty thousand years ago.

Modern Humans out of Africa

The emergence of *Homo sapiens*, anatomically modern humans, from 200,000 years ago, follows a pattern very similar to that set by *H. heidelbergensis* (and *H. erectus* even earlier), with initial evolution taking place in Africa, followed by migration to the rest of the world. The oldest well-characterized fossils of anatomically modern humans come from the Kibish formation in South Ethiopia, and their estimated date is 195,000 years ago, plus or minus 5,000 years.[4] Other well-established fossil skulls of our species have been found in the village of Herto in Ethiopia and date from 160,000 years ago as established by argon isotope dating.[5] Some limited expansion of our species had already taken place as far as the Levant by 115,000 years ago, as established by partial skeletons of unequivocal *H. sapiens* found at Skhul and Qafzeh in Israel. But significant emigration out of Africa does not seem to have taken place until after 70,000 years ago, with modern humans reaching right across Asia and on to Australia by 50,000 years ago, then backtracking into Europe by 40,000 years ago, where they are known as the Cro-Magnon people. By 15,000 years ago, they were trickling down into North America across the Bering Strait.[6]

Genetics has been of enormous help in tracking the evolution and migration of *H. sapiens*, whereas apart from Neanderthals and the finger sample from the Siberian cave, the remains of other hominin species are simply too old for extracting DNA samples. Mitochondrial DNA, with its 16,569 nucleotide base pairs and 37 genes, has been particularly useful in this respect for two distinct reasons.

First, its mutation rate acts like a genetic clock in that the accumulation of mutations can be roughly calibrated with time. The same is true for nuclear DNA, but because the mitochondrial mutation rate is about five to ten times faster than that of nuclear DNA, the mitochondrial clock gives finer resolution of recent events. The accumulation of mutations can also be used to gauge relative relatedness among different mitochondrial DNA sequences without the need for calibrating a genetic clock.

Second, mitochondrial DNA is inherited only from the mother, as previously mentioned, since the mitochondria coming from the sperm are lost from the zygote shortly after fertilization, leaving the mother's mitochondria as the source of all the mitochondria that eventually end up in all the cells of our bodies. So unlike nuclear DNA, which is constantly being mixed by the process of sexual reproduction, mitochondrial DNA provides a more straightforward reading of the mutational changes that accumulate over generations and that characterize variation between different human populations.

Simply by drawing family trees on the back of an envelope, we can easily show that the mitochondrial DNA of all the people alive in the world today must have originated from a single woman, the so-called Mitochondrial Eve (Figure 7.2). In some ways the term "Mitochondrial Eve," much used in the popular press when the concept was first introduced in the 1980s, is a misleading one. It does not at all mean that this woman was the only female human alive at her time or that she is our oldest female ancestor; it just means that if we follow the maternal lines in a population back through a large enough number of generations, we will eventually arrive at a single woman, the source of all mitochondria in the present human population. Let us imagine that there were 1,000 females alive at the time; the descendants of 999 of them eventually did not have any female children, perhaps thousands of years later, so their mitochondrial DNA was not transmitted. In this way the descendants of Mitochondrial Eve generated the present world population.

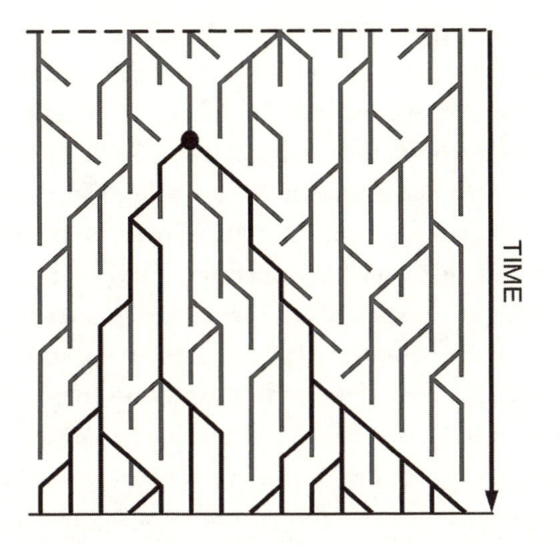

FIGURE 7.2. The most recent common female ancestor of the entire population of present-day humans has been dubbed "Mitochondrial Eve." The figure illustrates coalescence, the pattern of descent that means that a single most recent common female ancestor (the black circle) must exist for any extant population of sexually breeding organisms. Note that many other human females were alive at the time of "Mitochondrial Eve," but eventually they left only sons or their lineages died out. Also note that only mitochondrial DNA is inherited directly from "Mitochondrial Eve"; nuclear DNA is inherited in a much more complex pattern from multiple ancestors by sexual reproduction.

Extensive investigations of mitochondrial DNA sequences from human populations around the world have revealed two fascinating insights into the evolution of our species. For starters, the genetic clock places Mitochondrial Eve in Africa one hundred thousand to two hundred thousand years ago, consistent with the fossil data for the emergence of *H. sapiens* described above.[7] This finding provides strong support for the "Out of Africa" model for human evolution that currently holds sway.[8] Had there, for example, been extensive interbreeding between the *H. sapiens* populations and the more ancient *H. erectus* populations that overlapped extensively in time with *H. sapiens*, then much older dates for the putative Mitochondrial Eve would have been expected.

Another important observation arising from human mitochondrial DNA sequence studies is that fewer differences are present among all the non-African mitochondrial DNA in the world than is found among the mitochondrial DNA of different African populations. This is consistent with the idea that different human populations lived in different parts of Africa over the period from two hundred thousand to one hundred thousand years ago, and that the wave of emigration from Africa that occurred after seventy thousand years ago did so in the form of a relatively small number of people, perhaps a few thousand or less, who then spread out to transmit their mitochondrial DNA to what is now the whole non-African population of the world. We are so used to large human populations today that we forget that for a significant portion of our early history the human population was really small, the genetic data suggesting that at times the total population during the time when we were all Africans contained only a few hundred breeding pairs. From a scientific perspective this is important because evolution tends to occur more rapidly in small populations, but we could so easily have been yet another extinct species.

Other genetic methods besides the study of mitochondrial DNA are also very useful in the investigation of human evolution. For example, because the Y chromosome is only passed on through males, DNA sequences from this chromosome can be analyzed as a kind of male equivalent of mitochondrial DNA analysis. Genetic analysis of Y chromosomes from populations around the world suggests that the male individual from whom all current-day Y chromosomes in the world are derived lived in Africa around fifty thousand to one hundred thousand years ago.[9] Again remember that this does not mean that he was the only male alive at the time, only that his contemporaries failed to produce an unbroken male lineage until the present day, in much the same way that Figure 7.2 illustrates for female mitochondrial DNA transmission. Also the male from whom all our (males') Y chromosomes are inherited does not mark the time when our species emerged, only the time when our most recent male ancestor existed.

Although little data support the idea of significant interbreeding between anatomically modern humans and more ancient human populations, some data do point in this direction. For example, in addition to the Denisova gene flow to Melanesians already mentioned, the Neanderthal genome sequence results revealed the provocative finding that non-African humans are genetically closer to Neanderthals than African humans. In fact, the European and Asian genomes that were sequenced appear to contain 1 to 4 percent DNA of Neanderthal origin, and the gene flow that occurred appears to have been almost entirely from Neanderthal to human, rather than vice versa. How come? The most likely scenario is that there were a few instances of sexual reproduction between Neanderthals and human individuals belonging to the population that is thought to have emigrated out of Africa to populate the world sometime after seventy thousand years ago, explaining why the Neanderthal DNA sequences are not found in African genomes. The contribution of the Neanderthal genome has remained in European and Asian populations ever since. To put this in perspective, most of our genes are very similar anyway to those found in Neanderthals and chimpanzees, and to other mammals like mice. We all share a "how-to-build-a-mammal" instruction manual, and the relatively minor genetic differences between us (minor relative to those we share in common) are the icing on the cake, as it were, that make us a human rather than a mouse, a chimp, or a Neanderthal.

Our Common Inheritance with the Apes

How can we be so sure that we share a common inheritance with the apes? Some people feel uncomfortable with such an idea. Does this not threaten our notions of what it means to be a uniquely valuable human being, even made in the image of God? We leave this latter question until the final chapter, but for the moment it is worth emphasizing that genetic data well establishes our common inheritance. Even if we had no fossil data available at all, we could still be certain, based on genomics, about our common inheritance with the apes.

A direct comparison between the human and chimpanzee genome sequences is informative.[10] About 35 million single-nucleotide changes separate us from our nearest living relatives, representing about 1 percent of the genome. In addition, around 5 million insertional or deletional differences ("indels") make 3 percent of each genome different from the other based on this criterion. So the deletions and insertions win over the point mutations in terms of generating a different genome, and together they yield a 4 percent difference in total between us and the chimps. The proteins encoded by genes are extremely conserved, with 29 percent being identical and the rest differing by only two amino acids on average, whereas fifty-three known or predicted human genes are missing or disabled in the chimpanzee genome.

The biggest sequence differences of all between us and the chimps have been noticed in our respective Y chromosomes.[11] Much of the difference is due to gene loss in the chimp and gene gain in the human. The chimp Y chromosome has only two-thirds as many distinct genes or gene families as the human Y chromosome and 53 percent fewer protein-coding genes compared to humans. Furthermore, about 30 percent of the chimp chromosome has been totally rearranged, so it can no longer be lined up with the human equivalent over this portion. This is quite different from the nonsex chromosomes, which can be lined up very precisely so that sometimes you hardly know whether you're looking at a human or a chimp chromosome. As David Page, the researcher who reported these findings, remarked, "If you're marching along the human chromosome 21, you might as well be marching along the chimp chromosome 21. It's like an unbroken piece of glass. But the relationship between the human and chimp Y chromosomes has been blown to pieces."[12]

Why the huge difference? It's probably a composite of several factors, including the fact that most of the characterized genes on the Y chromosome are involved in making sperm, and strong selective pressures are present on the fitness of sperm for obvious reasons. This pressure is even greater in chimps, where receptive females often mate with many males in one session, so the male with the best sperm has the highest chance of reproductive success. In addition, 95 percent of the Y chromosome does not undergo crossing over during meiosis since only 5 percent of the Y chromosome is homologous with the X chromosome. Y chromosomes thus have no way of renewing and replenishing lost genes by receiving segments of another chromosome, as is the case for the nonsex chromosomes. Overall, then, the Y chromosome tends to evolve more quickly than the other chromosomes.

Comparing the transcription patterns of genes between humans and chimpanzees is also informative. Although the techniques still need refining, it is already clear that at least 10 percent of the genes in the brains of the two species vary considerably in this respect.[13] In other words, showing that two sets of genes are very similar between two species is not sufficient. The really significant differences between species may be explained by the regulation of gene expression, affecting which mRNA and RNA gene products are produced, and the timing of their production. Thousands of regulatory gene sequences are present in the human and chimpanzee genomes, and some of them differ in sequence. Since they play a critical role during development, the evolution of such regulatory sequences has very likely had a major impact on the divergence of the human and ape lineages.[14]

The series of speciation events that have led to separation from our chimpanzee cousins in evolutionary history also very likely involve the *prdm9* gene already introduced in chapter 6. The gene is remarkably different in its sequence between human and chimp, much more so than for other genes. It also regulates the "hotspots" in the genome where recombination is far more likely to occur during meiosis, and because the human and chimp versions of *prdm9* are so different, it turns out that their "recombination hotspot" profiles are very different as well. By such mechanisms are reproductive barriers raised between different species.

Genetic Fossils and Common Ancestry

As far as our common ancestry with the apes is concerned, the best evidence comes not from known functional genes, but from the thousands of genetic fossils that litter our genomes that have no known function—the pseudogenes, to be explained here—and the transposons and retroviral insertions already introduced in chapter 4.

Pseudogenes are stretches of DNA that are so similar to functional genes in other ancestral organisms that there is no doubt where they came from, yet so full of mutations that they are functionally incapable of making protein, as can be shown experimentally in the laboratory. Mutations may be in the regulatory regions, so transcription is switched off, or within ORFs, disabling protein production. Cars that are not serviced eventually become dysfunctional. Genes that are in excess of requirements for an organism to flourish in a given ecological niche are no longer under the pressure of natural selection to conform, so they accumulate mutations and fall into disrepair.

An estimated 12,308 pseudogenes reside in the human genome,[15] although some assessments put the real number as high as 19,000, rivaling the number of functional genes in our genomes. The reason for the uncertainty is that some pseudogenes are identified as such, but then found later to be functional; more commonly, a pseudogene is in such tatters that its firm identification based on ancestry takes some careful sleuthing. In general the number of estimated pseudogenes tends to edge upward as more genomic investigation of this kind is performed. "Real biologists" who study gene function using real living materials can rather dismissively refer to such computer-based data mining as experiments carried out "in silico." However, the collaboration between the "functional biologists" and the computer experts has been incredibly fruitful, and their very respectable and important profession is now known as "bioinformatics."

Fortunately for evolutionary biologists of both flavors, hundreds of examples of well-established pseudogenes can be used to confirm or establish evolutionary lineages. One classic example also helps explain why sailors in previous centuries used to suffer from scurvy on long voyages. The reason is that their stocks of fruit ran out after a few weeks at sea, depriving them of the vitamin C needed to prevent scurvy. So why can't we make vitamin C ourselves, given that most other mammals make it? Mammals such as rats and mice have a gene called *gulo* that encodes an enzyme called L-gulono-γ-lactone oxidase (GLO—hence the name of the gene) which is needed to synthesize vitamin C. But a mutation entered into the primate lineage more than 40 million years ago, so we all now have only the nonfunctional pseudogene version of *gulo* and have to keep eating all those oranges to stay healthy. Figure 7.3 shows the reason.[16]

	80	90	100
Rat	GAGGTGCGCT	TCACCCGAGG	CGATGACA
Human	G*G*GGT*A*CGCT	TCACC*T*G _ G*A*	CGATGACA
Chimpanzee	G*G*G*C*T*A*CGCT	TCACC*T*G _ G*A*	CGATGACA
Orangutan	G*G*GGTGCGCT	TCACCC*A* _ G*A*	CGATGACA
Macaque	G*G*GGTGCGCT	TCACCC*A* _ *A*G	CGATGACA

Figure 7.3

The nucleotide sequence (nucleotide numbers 80–107) of a part of the gene possesses a particular mutation. The rat sequence makes a functional GLO enzyme, so rats don't need to eat oranges to stay healthy. The primate gene sequences are nearly identical to the rat's, but they lack an "A" at nucleotide 97. Such a mutation would have occurred originally as a random, highly unlikely event, and yet it is found in all the primate species examined. The most plausible interpretation is that the mutation arose in an ancestor of all the great apes and macaques, and all these species have inherited the mutation ever since. The chances of the same mutation happening more than once in exactly the same genetic letter in a gene thousands of letters long is vanishingly small. When one considers that hundreds of such pseudogenes can be used to establish our evolutionary inheritance with the apes, the data based on pseudogenes alone becomes overwhelming.

But that is not all. Transposons (the jumping genes introduced in chapter 4) are another type of genetic fossil that provide equally powerful data demonstrating our common inheritance. If one of these sequences has been inserted at a specific location in the genomes of different species, then this demonstrates unequivocally that these species must all have descended from the same common ancestor. We share nearly all (99 percent) of these 3 million fossilized inserts into our genomes with chimpanzees, most with macaques, and many with distantly related mammals. They are found, with very few exceptions, at precisely the same locations on the equivalent chromosomes in the different species—in other words, these sequences could not possibly have ended up at these precise spots in our own genomes by a random process independent of our ancestry. Conversely, in large surveys of human-specific transposons, independent insertions into the equivalent chromosomal sites of other species by transposons specific to those species have not been observed, as you might expect if something special about those particular sites favored such insertions.[17]

As discussed in chapter 4, an estimated 1.1 million "*Alu*" transposons alone exist, accounting for about 10 percent of our entire genome. Of these, nearly all the 1.1 million are shared with the chimp genome and 1 million with the macaque genome. Most were added to our genome about 45 million years ago when, it is estimated, approximately one new Alu insertion took place for every live birth. Taken together, the human lineage has seen around ten thousand further transposons added to its genome since our last common ancestor with the chimps, whereas the chimpanzee lineage has seen about five thousand such events,[18] representing less than 1 percent of all the transposons that are shared in common. This does not mean that all these events have occurred in either chimps or modern humans, but rather along the way in one of the intermediates—hominins in our case—that connect our respective lineages all the way back to our last common ancestor.

To give an idea of how powerful such observations are for understanding our ancestry, consider the example illustrated in Figure 7.4.[19]

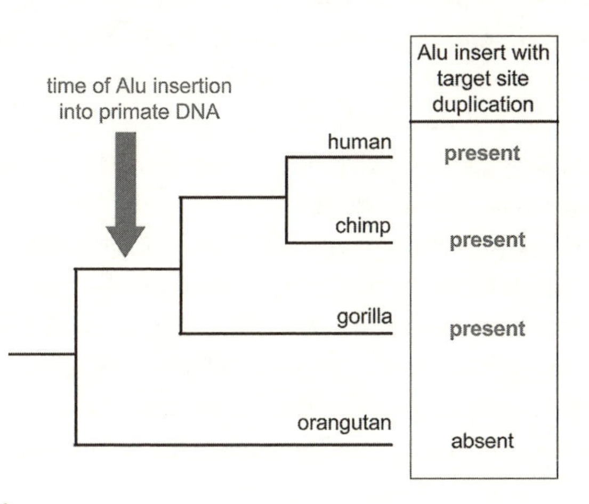

FIGURE 7.4

In this example, the jumping gene is an Alu insert. It is hundreds of nucleotide base-pairs long. The gene has inserted into a non-protein-coding region of the genome of the common ancestor of human, chimp, and gorilla, but is not present in the DNA of the orangutan. This tells us immediately that the insertional event must

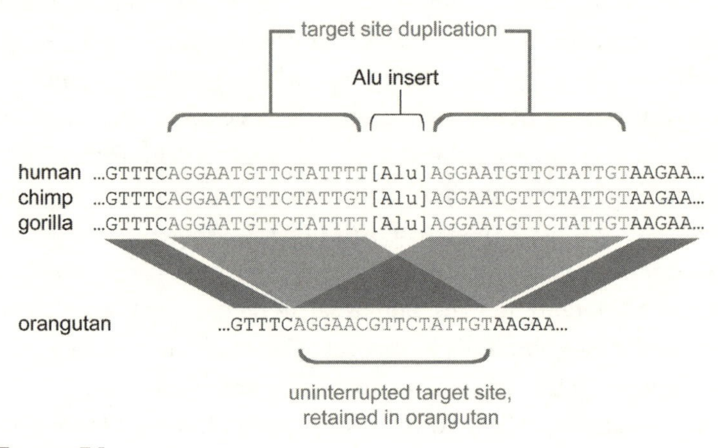

Figure 7.5

have taken place after our lineage split off from the orangutan about 12 million to 15 million years ago. Just to demonstrate how easy it is to identify the exact Alu insert, Figure 7.5 shows the precise insertion site in the genomes of the species involved.

Looking along the DNA letters (nucleotide bases) in the chromosomes of human, chimp, and gorilla, we can pick out immediately that the Alu insertion sequence has located in exactly the same place in each genome, with only tiny mutational changes between the sequences around the insertion site. But the insertion is completely missing from the orangutan at precisely the same spot where the uninterrupted sequence is shown. Multiply this type of observation thousands of times, and you get an idea how powerful these genetic fossils are in revealing our evolutionary past.[20]

As always in science, caution is required when handling such data. One reason is that, very rarely, an Alu insertion can be excised from a genome at a particular site on the chromosome, although nearly always a telltale sequence signature is left behind as a marker of where this happened.[21] If one was dependent on only a few transposons to map evolutionary lineages, then this might be a cause for concern, but because humans and chimpanzees share 3 million transposons, one is really spoiled for choice in picking a genetic fossil to use for investigations of primate evolution.

The retroviral insertions that constitute 8 percent of the genome are also useful in this respect, providing plenty more examples of genetic fossils. A single example serves to make the point, a human endogenous retrovirus (HERV) known as K105 (see Figure 7.6). This figure shows the insertion site of K105, thousands of nucleotide base-pairs long and indicated here as [HERV], which is located at precisely the same site in our own genomes and those of our nearest relatives.[22] Again, by tracking back in the primate record far enough in evolutionary time, we can locate the point at which K105 no longer appears in the genome. In other words, on a particular day of a particular week in a particular year, millions of years ago, that particular retroviral DNA sequence was inserted into a particular germ cell of one of our ancestors, and it has been there ever since; we all continue to make millions of copies of it, very precisely, every day of our lives. That is an amazing thought.

human	CTCTGGAATTC[HERV]GAATTCTATGT
chimpanzee	CTCTGGAATTC[HERV]GAATTCTATGT
bonobo	CTCTGGAATTC[HERV]GAATTCTATGT

Figure 7.6

Although we have mainly been interested here in the use of pseudogenes, transposons, and retroviral insertions as tracking devices for the construction and confirmation of evolutionary lineages, let us not forget that they have also contributed very directly to the evolution of the human genome in myriad ways. One way is, of course, in the sheer acquisition of bulk: without them our genomes would be more than 50 percent smaller. On average there is an Alu insertion every three thousand base-pairs, with one new insertion for every twenty live births in the human. Some people might call that "flab," but out of the flab has come some useful acquisitions. It is good to remember the retroviral origins of the syncytin-1 and syncytin-2 placental genes that we discussed in chapter 4.

In addition, these "DNA extras" have contributed to changes in genomic instability and gene expression, and have led to other types of gene innovation. Certainly not all these changes have been beneficial to the organism involved, to put it mildly. But other changes have been beneficial. Some pseudogenes mutate back to their original functions or acquire new ones. Increased genomic instability at sites of Alu insertion has increased the deletion of DNA that sometimes contains genes—on occasion, unwanted genes. It has been estimated that during primate evolution as many as forty-five thousand insertional events led to the removal of more than 30 million base-pairs of DNA. The same genomic instability has led to gene duplication events near transposons that in turn have generated whole new families of genes with novel properties.[23] Transposons, pseudogenes, and retroviral insertions together provide a great big stir of the genomic stew that keeps everything moving onward to more novelty and creativity. Genomes never stand still, especially with this kind of help from their friends.

CHAPTER 8

The Rainbow Diversity of Humanity

THE VERY FIRST almost complete human genome sequence published in 2005 was a composite of DNA from five different anonymous individuals. When I went for a wander around the Cambridge Sanger Centre's human genome sequencing lab in the 1990s during an early stage of the project's development, I asked my guide where they got their DNA from for sequencing. "Oh, we just bleed each other," he replied cheerfully. For the final sequence the process became more formalized to ensure anonymity and prevent any subsequent lawsuits. Because the first complete sequence was a haploid version, the differences between paired chromosomes were not addressed.

Life has changed dramatically since those early days. At first it was just celebrities who had their genomes sequenced: a Nobel Prize–winner, a CEO of a genome company, and a bishop. Then the emphasis shifted to representative individuals from different ethnic backgrounds. At the time of this writing, a few hundred human genome sequences have been completed. By the time this book appears, the numbers will be shooting up into the tens of thousands. Eventually it is very likely that all babies born will have their complete DNA sequence stored with their medical records (hopefully securely).

The main reasons for this dramatic explosion of human genome sequence information are mainly the advent of new technologies with a striking decrease in costs, together with the drive to understand the contribution of genetic diversity to disease susceptibility, diagnosis, prognosis, and treatment. The Human Genome Project that culminated in the first complete sequence took more than a decade of effort and $3 billion to accomplish. The second complete sequence cost $100 million, and the third, that of James Watson, $1.5 million. Today a human genome can be sequenced in a day on a single machine at a cost of a few thousand dollars or less.[1] The cost of sequencing a million DNA base-pairs dropped from five dollars in 2008, to two dollars in 2009, to less than one dollar by late 2010. A $10 million Archon X-Prize is on offer for the first team that can build a device that can sequence one hundred human genomes to high accuracy at minimal cost within ten days. In 1989, when the Human Genome Project really took off, such a goal would have seemed way beyond reach. At the time the best machines were sequencing just forty-eight hundred base-pairs a day. Today machines can already sequence 100 billion base-pairs per day and will soon be able to churn out three times that level. Of course, the sequence data itself still has to be processed to construct the precise sequence for each chromosome, using incredibly high-powered computing techniques, but this has now become pretty routine.

It is not just human genomes that are benefitting from these great advances in technology. More than four thousand genomes have now been sequenced from nonhuman species. Although the bulk of these are viruses, bacteria, and bacteriophage (viruses that infect bacteria), organisms that are relatively easy to sequence, they also include dozens of more complex plants and animals, including a good collection of mammals. Eventually it seems very likely that representatives from each species will be routinely sequenced, greatly helping in sorting out taxonomic and evolutionary relationships in the great bush of life. There is still plenty of discussion about which twig goes exactly where, or even the precise location of larger branches.

The first big surprise that came in the era of personalized genomes was when Craig Venter's paired chromosomes were sequenced separately to generate the first complete diploid sequence data.[2] Venter is the CEO example alluded to above, founder and president of the J. Craig Venter Institute in La Jolla, California. The surprise came in finding out that the differences in the DNA sequences between his two chromosomes were greater than imagined—in fact, in the region of 0.5 percent considered altogether.

To gain some appreciation of differences between chromosomes, the extent of human variation in the genome taken as a whole is first worth considering. Clearly variation can only be measured if there is some gold standard against which differences can be contrasted, and this is provided by the composite "reference human genome sequence" arising from the Human Genome Project led by Francis Collins, the first complete, publicly funded sequence to be published.[3] With that in mind, Venter's diploid sequence revealed more than 4.1 million DNA variants, in all the shapes and sizes previously surveyed in chapter 4, when compared to the gold standard. These included 3,213,401 single nucleotide polymorphisms (SNPs), the point mutations that affect only a single letter in the genetic alphabet. In addition, there were 53,823 substitutions of small sections of DNA from 2 to 206 base-pairs in length; 292,102 heterozygous insertion/deletion events (indels) that occurred on only one chromosome at a particular site, ranging in length from 1 to 571 base-pairs; 559,473 homozygous indels, ranging in length from 1 to 82,711 base-pairs that occurred on both chromosomes of a pair; and 90 chromosomal inversions. As a result of all this variation, 44 percent of the known protein-coding genes have at least one, and often more, variation in their sequence. The reason for giving all these numbers (which may call for some review of the material in earlier chapters to follow properly) is simply to point out that the range of different types of variation in a single human genome is truly impressive.

From the numbers given, it might at first appear that the SNPs represent by far the greatest type of variation. Although correct numerically, this is not the case once the actual length of the variant DNA is measured in base-pairs. Using this as a measure, it turns out that SNPs account for only 26 percent of the variant base-pairs, whereas the dominant 74 percent are brought about by all the other types of variation.

We are now in a more informed position to look at that 0.5 percent variation between Venter's chromosome pairs. What does it consist of? Around 52 percent of the SNPs were heterozygous, meaning that these were single nucleotide base differences found on one chromosome but not the other, the other 48 percent of the SNPs being homozygous (i.e., found on both chromosomes). From the numbers given above, we also notice that nearly 300,000 of the indels were heterozygous. Taken together, these indels account for much of the 0.5 percent variation. Each chromosome is a mosaic composed of a different family history, converging in a single individual to create a unique human being.

Further sequencing of diploid human genomes, together with a worldwide investigation to uncover genetic diversity in people from a wide range of ethnic origins, has continued to uncover an impressive collection of differences. Known SNPs in the human population now number around 18 million,[4] although rare variants probably run into the billions. Differences in "copy number variation" also vary considerably. This terminology is somewhat confusing because we have already referred (in chapter 4) to the fact that copy number of actual genes can vary between individuals. However, this is not what "copy number variation" (often abbreviated to CNV) in this context refers to, but rather the random duplication of small segments of the genome, or tandem arrays of such repeats all in a row, or even deletions of such larger segments. We are talking here about a few hundred base-pairs or more, and the CNV terminology therefore encompasses much of the chromosomal structural variation that was discussed in chapter 4.

In one study, for example, DNA from 450 individuals of European, African, or East Asian ancestry was investigated for the presence of CNVs, and no fewer than 11,700 variable regions of 443 base-pairs in length or longer were uncovered in this cohort, representing 3.7 percent of the whole genome.[5] The greater variation in the African samples from this study is consistent with the much longer time that human evolution has been taking place in Africa compared to the rest of the world.

As already mentioned, in the current "Out of Africa" model for the emergence of the present world's population of modern humans, a narrow genetic bottleneck occurred as the great trek began out of Africa from seventy thousand years onwards or later. The effective population size of the emigrant population has been estimated at between 60 and 1,220 individuals,[6] meaning that virtually all the world's present non-African populations are descended from this tiny founder population. Even the bugs inside human guts tell the same story with their genetic variation reflecting the African origins of their hosts.[7] But within Africa different groups of humans were living for at least 130,000 years before the emigration, many of them isolated from each other for long periods of time. Therefore, one would expect greater genetic variation between different populations of Africans than between different populations of non-Africans, which is in fact what is observed.

The same point is also valid for variation in the SNPs. One of the early human diploid genomes to be sequenced was that of a male Yoruba from Ibadan in Nigeria.[8] This person's genome has 3.8 million SNPs, of which 2.4 million (63 percent) are heterozygous and 1.4 million (37 percent) are homozygous. Compare this with Venter's European-descent SNP heterozygous level of 52 percent mentioned above. In general, the level of SNPs is about a third more in African DNA than in European, again consistent with the inference that all non-Africans from the world originate from a relatively recent and rather small population.

Taken overall it appears that we all differ from each other by around 0.5 percent of our genomes, although this figure could range as high as 2 percent. It has been estimated, based on current findings, that if all the human genomes of the world were compared, then somewhere in the range of 19 million to 40 million base-pairs in total would vary between individuals.[9] This is a dramatic increase when we consider that in 2001 we thought that this interindividual difference was only 0.1 percent, but only SNPs information was available in the first draft of the genome, and fewer SNPs were known then than now. In practice, if we invited guests for tea from, let's say, 140 different countries, from every continent of the world, then everyone in the room would differ from each other by about 1 out of every 1,000 base-pairs, and by about 1,300 CNVs in total, together with some of the rarer differences like individual gene copy numbers mentioned above. Common variants account for most (90 percent) of the differences between any two individuals.

Given that the genetic difference that separates us from the chimps is only 4 percent, that 0.5 percent to 2 percent variation sounds quite a lot. But two points are important to remember. The first is: Which chimp? The chimpanzee genome sequence was derived from a single male individual. But chimps display interindividual variation just as we do, or indeed as any individual within a species does. Not surprisingly, we know a lot less about genetic variation between chimps than we do about variation among humans. So the 4 percent genetic distance between us and chimps should be treated as an initial rough estimate only.

The second point arises from the fact that most of the interindividual genetic differences between humans make no difference at all, because most differences are found in the 95 percent of the genome that has no known function, or maybe just no function whatsoever. Even considering the fact that noncoding regions of the genome are transcribed into mRNA, as discussed in chapter 2, this in itself doesn't mean that the DNA segments involved are functional. Even within the protein-coding genes, RNA genes, and regulatory sequences that together make up about 5 percent of the genome, many SNPs make no difference because they represent conservative changes: the amino acid sequence of the encoded protein remains the same. In fact, any individual typically differs from the reference genome at around 11,000 sites in the genome that make a difference to the amino acid sequence of proteins, whereas changes at a further 11,000 sites make no difference. On average, each person carries 250 to 300 variant genes that are expected to make a difference to gene function,[10] although many of these are heterozygous rather than homozygous and so their effects may well be masked by use of the back-up version of the gene on the second chromosome.

Buried within those thousands of genetic differences that distinguish us from each other lies an incredibly important cohort that, among many other aspects of our being, defines our blue eyes, knobbly knees, height, and short- or long-sightedness, and in addition makes important contributions to our personalities. The genetic variation that encodes such differences may be mediated through regulatory sequences, differences in transcription factors, differences in the timing and expression levels of genes in different tissues, changes in alternative splicing, changes in protein amino acid sequence, differences in RNA gene sequences, and much else besides.

How can we find these functional genetic differences in the genomic haystack? Diseases or disease susceptibilities often point the way, as we discuss in the next chapter. For the moment we should note that our uniqueness as individual human beings is strongly underpinned by this great array of interindividual genetic variation. Without it we—humanity—would comprise one giant clone, and we would all look identical. Life would indeed be very boring.

Race, Migrations, and Genetics

When the South African civil rights activist Archbishop Desmond Tutu agreed to have his DNA sequenced and published in a top scientific journal, he knew full well what he was getting into.[11] Tutu is from the Bantu, the majority people of South Africa (Figure 8.1). The purpose of the study was not to see whether anything distinctive resided in the genome sequence of an archbishop, but to compare a Bantu genome with a genome from the ancient and diverse population known as Khoisans (from the Khoi and San peoples).

FIGURE 8.1. The map shows the localities of four Khoisans (a–d) and one Bantu (e) from four different language groups: a) Tuu, b and c) Ju/'hoansi, d) !Kung (Etosha), and e) Xhosa/Tswana. Although they exist in close geographic proximity, there is more genetic variation among the genomes of these Africans than among the genomes of Europeans and Asians. Adapted by permission from Macmillan Publishers, Ltd. Schuster, S., et al. *Nature* 463 (2010): 943–47.

By comparing the complete sequence of one Khoisan with the partial sequence of three others, it was possible to show that they were genetically more different from each other than a European is from an Asian. The mitochondrial DNA sequences from Europeans show, on average, 20 base-pair differences from the reference sequence, whereas the South African sequences showed up to 120 base-pair differences from the same reference. Even more striking is the fact that up to 84 differences were observed in pair-wise comparisons between the different South African genomes. Once again, prolonged separation in different tribes over tens of thousands of years holds the clue. A separate repertoire of SNPs and other variations will continue to accumulate as long as little or no gene flow mingles the sets of variants together.

As far as the Khoisan tribesman and Archbishop Tutu are concerned, each carried around 1 million SNPs not found in each other, nor in other published human genome sequences, including that from the Yoruba individual from West Africa. The SNPs were not distributed evenly across the genome but rather in a number of striking hotspots in which variation occurred with much greater frequency. One hundred and ninety-three genes differed in copy-number between the two genomes. The Khoisan had no fewer than fifteen copies of the salivary amylase gene, consistent with a forager lifestyle in which starch has likely played a significant role in nutrition. Other features of the Khoisan genome data reflect their tough desert life. Three of the Khoisan carry a version of the ACTN3 gene expressed in muscle that is thought to lead to faster sprinting. All have a taste receptor gene that enables certain plants to taste bitter, preventing them from eating plants that might prove toxic. But they lack a gene variant found in other African populations that would give resistance to malaria, more often found in agricultural populations where mosquitoes are endemic.

Are these kinds of comparisons between the genetics of different ethnic groups potentially "racist"? It is difficult to know why this should be the case. Of course, the link between race and genetics has had a fraught history, not least through the horrors of twentieth-century eugenics and the horrific policies of the Third Reich.[12] But these gross aberrations should not lead us into the opposite extreme of denying that genetic differences between human populations make any difference at all. Clearly they do. For a start, we look different. If a population lives in relative reproductive isolation for tens of thousands of years, then it may accumulate different sets of genetic variants from other populations. Many of these will be due to random genetic variation, but other genes will be under natural selection, helping people to adapt to their particular environments. Nevertheless, no sets of genetic variants identify discrete boundaries between "races"; also the variation between human populations is a small fraction of the total variation found among humans. Far more variation is found among individuals within a population in a particular geographical location than the average difference between that population and another one elsewhere. In fact, only 15 percent of total human genetic variation can be assigned to the differences that divide populations living in different continents, whereas 85 percent of the variation represents the average difference between different members of the same population.[13] This is well illustrated by the fact that Watson and Venter's genome sequences share more SNPs with a Korean individual than they do with each other. This doesn't mean that this would be true of a comparison with all Korean genome sequences, just that in this particular case the 15 percent of intercontinental differences were swamped by the 85 percent of interindividual differences.

But if we just focus on the 15 percent for a moment, then the adaptive responses of populations to different environments is there for all to see. The skins of humans became whiter as they moved from the tropics to cooler environments. An allele of the pigmentation gene SLC24AS that contributes to light pigmentation is present in almost all Europeans but is nearly absent in East Asians and Africans. Humans such as Inuits and Lapps who have lived for thousands of years in Arctic conditions have developed stocky bodies that lose much less heat than the tall Africans of the tropics. Specifically, they have shorter shins relative to their thighs, an adaptation to the cold that also evolved in the Neanderthals as they coped with the ice ages of Europe.

Yet none of these adaptations, useful as they are, go anywhere near providing a biological basis for the notion of race as constructed, for example, over the past few centuries in Western cultures. Linnaeus came up with a classification of 6 "races" in 1735, the Chicago Natural History Museum with 105 in 1933, and the Metropolitan Police Service in London with 16 in 2005, different from the list that had been used before.[14] But none of these racial classification systems can be identified with any discrete set of genetic variations; arguably the different classification systems are related more to the economic and cultural requirements, as well as prejudices, of those generating them. Certainly from a biological perspective, the genetics of skin color is not where one would begin if seeking to establish some kind of racial classification system: an estimated seventy different variant genes affect skin pigmentation. Once we realize that genetics provides no basis for the separation of humankind into distinct racial types, we can then relax a bit and appreciate the rainbow variation that makes us as a species as different as we are.

For medical reasons the International HapMap project aims to map the diversity that exists between individuals all around the world.[15] To take DNA samples, all that's necessary is to swill your mouth around and spit into a clean receptacle—it is that simple, as every reader of crime novels knows by now! The particular set of nucleotide variants on a particular chromosome is known as the "haplotype," hence the abbreviation "Hap" in "HapMap." A fascinating consideration is that genetic variation is not simply a reflection of a long-lost past ancestry but reflects relatively recent natural selection as well.

Bioinformatics can be useful in this respect in identifying genes that have undergone recent natural selection—for example, those that help protect against pathogens. Variants of two genes that protect against infection with the Lassa virus are favored in West Africans.[16] The reasons for the prevalence of those variants in that particular area are obvious enough, just as the prevalence of the hemoglobin S gene in areas where malaria is common can readily be explained by the resistance it provides to that debilitating disease, as we discussed in chapter 5. In other cases genes are identified as having undergone strong recent natural selection, but the phenotypic advantage they are assumed to provide is not so obvious.[17]

For example, a set of genes encoding the NRG signaling pathway is known to be important in the development of the heart, nerves, and breasts. Variants in these genes have been associated with schizophrenia and other psychiatric conditions. A particular set of variants has been selected for in non-African populations, but why this should be the case is a mystery. Sometimes the very fact that genes display evidence of recent natural selection can lead researchers to take a particular interest in them, occasionally uncovering medically significant results in the process.

Such reflections lead to the obvious question: is human evolution still in progress? In one sense the answer is "yes." Some genes are undergoing natural selection as we have just seen. Cohorts of SNPs vary between people who have lived in different parts of the world for extended periods, to the extent that a SNP fingerprint can be used to identify origins and migrations. This is evolution in process. But this is a very long way from the claim that eventually humans might evolve into something quite different, such as a new species altogether, and several good biological reasons explain why such evolution is very unlikely.

The first relates to the benefits of Western hygiene and medical care that have spread around the world, much of this as a direct result of medical missionary work that gathered pace from the nineteenth century onward. The chances of people reaching reproductive age, even if afflicted by different diseases or other disabilities, have vastly increased in the last century, thus slowing down or even stopping dead in its tracks any significant human evolution. Modern medical care is a great genetic leveller, giving people a greater opportunity to achieve reproductive success as its benefits become more widely distributed. Clearly we are not yet at the point where the playing field is equally level for all, but there is at least partial success.

Second, as human populations increase in prosperity, they tend to have fewer children for social reasons. The criterion of "differential reproductive success" is clearly less significant when the difference in reproductive rate is between the "replacement rate" of 2.1 offspring per couple compared to 6 to 15 children per couple.

Third, speciation occurs in isolated populations, but in today's global village, the existence of such an isolated population is highly unlikely, especially given that speciation in mammalian populations most likely takes tens of thousands of years.

The possible speciation events that might occur in human populations all arise from sci-fi scenarios. The first is the colonization of a distant planet by a human population that remains isolated from planet Earth for tens of thousands of years—a most unlikely scenario, at least for the time being. The second is a global catastrophe that results in a severe shrinkage of the total human population and a loss of the technologies that make possible the three factors mentioned above (medical care, high standards of living, and the global village). Plenty of apocalyptic novels explore this scenario, but its actual likelihood is hard to predict. The third is the deliberate changing of the genomes of certain individuals, but not others, by genetic engineering, to create a new race of beings that could reproduce only among themselves. Apart from the fact that such genetic manipulation is currently scientifically impossible—and illegal—such developments would be likely to occur only if some mad dictator decided to enforce such an enterprise, and one hopes and prays that such a horrendous initiative never materializes.

So overall there seem to be many good reasons for celebrating human genetic diversity.[18] It ensures that we all have unique genetic profiles, underpinning our distinctive individual identities. Identical twins have nearly identical genomes, but the epigenetic modifications to be considered in chapter 10, together with mutations accumulating in their somatic cells with age, means that identical twins are not really identical even at the genetic level, let alone at the level of their own personal choices and environments.

Genetic diversity is also important for the general health and viability of the human genome taken as a collective whole. Beneficial genetic variants that evolve in one population may be passed on to other populations by gene flow. Increased resistance to disease that develops in one area can be shared with the rest of humankind as the mingling of populations proceeds over the generations.

Human genetic profiles can also be very useful from a practical point of view in understanding and treating disease, not least in predicting which medication to give to which patient to avoid unwanted side effects, as we consider in the next chapter. A further practical application comes in tracking ethnic roots and relationships.

Genetic Variation and Ancestry

What's in a human hair? The answer for genetics is "a lot." Hair is one of the best sources of uncontaminated DNA—of special value when it belongs to an ancient human, and even better when it's been kept frozen. In 2008 the Danish biologist Eske Willerslev found a great treasure lying unwanted in the basement of the Natural History Museum in Denmark: a tuft of hair that had been extracted from the Greenland permafrost some twenty years earlier and kept in storage. From this modest amount of material, Willerslev and his team were able to derive the first genome sequence (79 percent complete) from an ancient human, an Inuit who lived about four thousand years ago.[19] By comparing the sequence with that observed in modern human populations living in the Arctic, it was possible to deduce that the sequence was from a male member of the Arctic Saqqaq, the first known culture to settle in Greenland (Figure 8.2). The team also showed that the Saqqaq are closest genetically to the Chukchis from Siberia, and that beginning about fifty-five hundred years ago, this man's ancestors migrated from Siberia, across the Bering Strait, then east through Alaska and Canada to Greenland, settling there about fifteen hundred years later. From the 350,000 SNPs detected (of which 7 percent were not previously reported) in the ancient genome, it was also possible

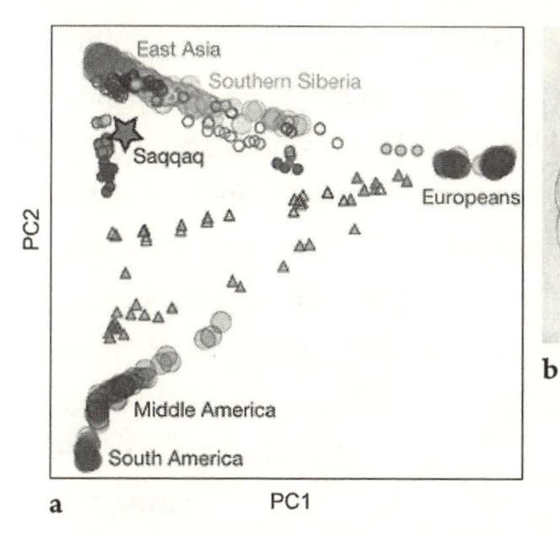

a

b

FIGURE 8.2. a) Principal component analysis (PCA) of the nuclear genome from the remains of a four-thousand-year-old male Paleo-Inuit along with genomes representing thirty-five populations of modern humans. This type of mathematical analysis reveals that the genome of the Paleo-Inuit is most closely related to the Saqqaq, the first known settlers of Greenland, who are descended from the Chukchi of Siberia. Adapted by permission from Macmillan Publishers Ltd. Rasmussen, M., et al. *Nature* 463 (2010): 757–62. b) A reconstruction of the ancient human based on genomic information suggests that he was balding, dark-skinned, and brown-eyed in addition to having dry-type earwax. Reprinted by permission from Nuka K. Godtfredsen.

to infer that the Inuit "had an A+ blood group, brown eyes, non-white skin, thick dark hair and 'shovel-graded' front teeth typical of Asian and Native American populations. What's more, he had an increased susceptibility to baldness, dry earwax and a metabolism and body-mass index commonly found in those who live in cold climates."[20] And all that from a single clump of hair!

If all that information can be obtained from the DNA extracted from hair that's four thousand years old, then it's no surprise to find that intensive investigation of contemporary ancestry and population studies has become something of a boom industry. As sequencing costs continue to decline, this boom is set to continue.

A highly informative SNP analysis of twenty-five diverse groups in India illustrates this trend.[21] The results revealed that India has two ancient populations, genetically divergent, that are ancestral to most Indians today. One, the "Ancestral North Indians," is genetically close to Middle Easterners, Central Asians, and Europeans, whereas the other, the "Ancestral South Indians," is as distinct from the Ancestral North Indians and East Asians as they are from each other. The Ancestral North Indian profile constitutes 39 to 71 percent of most Indian groups, and is higher in traditionally upper-caste and Indo-European speakers. The indigenous Andamans Islanders are unique in being Ancestral South Indian in origin without any ancestry from the North at all. Allele frequency differences between groups are larger in India than in Europe, reflecting the fact that different populations within India were originally founded by a relatively small number of individuals whose genetic fingerprint has been preserved over thousands of years.

The language of our genetic diversity can tell us a huge amount about our personal family histories, the histories of our ethnic groups, and the migrations of populations over the millennia. It is also very useful for catching criminals.

Forensic DNA: Liberation of the Innocent

In 1987 the Florida rapist Tommy Lee Andrews became the first person in the United States to be convicted using DNA evidence after raping a woman during a burglary. He was sentenced to twenty-two years in prison. The very first use of DNA in a court case had been the previous year when Richard Buckland confessed to the rape and murder of a teenager near Leicester in the United Kingdom, but his confession was not accepted by the court because his DNA did not match that of the semen left at the crime scene.

How certain can we be that so-called forensic DNA reliably identifies one individual and not another? The term refers just to plain ordinary DNA—the adjective "forensic" refers to its use, not to anything special about the DNA itself. The DNA used for analysis may come from a coffee cup, a razor blade, a single human hair, or a trace of semen.

The use of DNA for crime-busting was first pioneered at Leicester University by Professor Alec Jeffreys, who was later knighted for his pioneering work. In its contemporary form, the technique depends on the presence of short, repeating sequences of DNA scattered throughout the genome, called short tandem repeats (STRs). These are generally just four (but may be three or five) nucleotides long, representing yet another layer of variation between individual genomes in addition to those we have surveyed so far. In addition to their variation in length, the absolute numbers of STRs also vary widely, so that by using a range of ten to seventeen STRs, we can obtain a genetic fingerprint that is not unique, but which is possessed by only a small number of people in the world.

How few? It all depends. For example, there are about 12 million monozygotic (identical) twins on the earth, so they will all have identical DNA fingerprints anyway. It will also depend on exactly how many STRs are analyzed. The CODIS system used in the United States depends upon thirteen different STRs, which means that the chance of someone else having the identical DNA fingerprint is very small, but "very small" does not mean impossible. If you toss a coin enough times, then sooner or later you will get twenty heads in a row. Searching for a match based on the analysis of thirteen different STRs, investigators found a perfect match after searching only thirty thousand DNA sample sequences in Maryland in 2007. That does not mean that the same genetic fingerprint would be found after searching through any random thirty thousand DNA samples obtained from different states. It just means that caution is required.

In fact, one of the most powerful uses of forensic DNA evidence is to *exclude* someone from suspicion. In December 2005 Evan Simmons was proven innocent of a 1981 attack on an Atlanta woman after serving twenty-four years in prison. More than 165 individuals have been freed in the USA using postconviction DNA testing, and many in other countries also. By March 2009 Sean Hodgson had spent twenty-seven years in jail, convicted of killing Teresa de

Simone, age twenty-two, in her car in Southampton, in the United Kingdom, thirty years ago. But the conviction was overturned because DNA fingerprinting demonstrated that DNA found at the scene of the crime was not his. In many ways the innocent should be most grateful for the benefits of DNA testing. Had the technique been in wide use decades ago, then equally many innocent people might have been saved the anguish of languishing in jail over a period of many years or even decades.

Forensic DNA techniques have also been used to identify the victims of the September 11, 2001, bombing of the World Trade Center in New York. Forensic data have been used to track down the bodies of some of the thousands of activists who disappeared during Argentina's military dictatorship of the 1970s and early 1980s and their children, who were consigned to orphanages or illegally adopted by military families. A number of those children have now been reunited with their biological families. Chinese police have used forensic DNA to help hundreds of families reunite with their abducted children. In the 1950s Anna Anderson claimed that she was Grand Duchess Anastasia Nikolaevna of Russia; in the 1980s, after her death, her tissues that had been stored at a Charlottesville, Virginia, hospital following a medical procedure were DNA fingerprinted, revealing that she bore no relation to the Romanovs.

The sheer breadth of the technical applications of forensic DNA is extraordinary, and there is little doubt that more is yet to come, especially as ever more sensitive techniques come into play. We can celebrate our genomic diversity not just because of the underpinning that it gives to our human identity but also because of its amazing range of applications, from the tracking of human migrations, to the elucidation of ancestral roots, to the conviction of criminals and the freeing of the innocent from prison.

But we now need to turn to the darker face of genetic diversity: genetic variation as the cause of disease.

CHAPTER 9

The Genetic Basis of Disease

WHEN PAUL QUINTON was a kid, he used to cough a lot, and his sweat was so salty that it made his wire clothes hangers rusty.[1] Later on, when he was a student at the University of Texas in Austin, and his thoughts turned to marriage, he thought that he had better find out what was wrong with him. So Quinton went to the university library and found a medical description that fit his condition perfectly: cystic fibrosis. Aged nineteen he was suffering from a disease that normally would have killed him, although little was known about the underlying mechanism of the disease at the time (the average life expectancy of a baby born with cystic fibrosis in 1959 was only six months). In Quinton's case it led him into a lifelong research career to find out what was really happening, an ongoing saga to which we return in a moment.

We have already encountered cystic fibrosis in chapter 1 as an example of a disease illustrating recessive inheritance. Carriers bearing a single copy of the defective gene have no disease at all; the disease only occurs when both copies of the mutant gene are present in the homozygous condition. We should be very happy about that since we all carry fifty to one hundred variant genes implicated in various genetic disorders, but because we are generally heterozygous with respect to these variants, most are benign.

The first phase of discovery of inherited genetic diseases focused almost entirely on medical conditions such as cystic fibrosis, diseases revealed as being inherited in a recessive, dominant, or X-linked pattern (see chapter 1). Indeed this phase goes all the way back to the early decades of the twentieth century when conditions such as alcaptonuria were first discovered. Most often the familial pattern of inheritance is recognized first, which then leads to a hunt for the defective gene.

Today more than six thousand genetically inherited diseases have been described, and more are being identified all the time. The reason that you may read different estimates of the numbers of genetic diseases is because it all depends on what you include in the list. The best way to keep up to date is via websites such as the Online Mendelian Inheritance in Man site hosted by Johns Hopkins University.[2] This site shows 2,758 conditions for which the underlying mechanism is known, and another 1,780 conditions for which the Mendelian inheritance is established, but where the mechanism remains unknown.[3] Another useful website is the Human Gene Mutation Database, which reports that more than 70,000 different mutations known so far are involved in human disease.[4] Some genetic diseases can be subdivided into numerous subtypes, further increasing the numbers on the list. New entries to the online list (each one not necessarily representing a new disease) are running at more than 700 per year, whereas additions and corrections to entries already present amount to more than 10,000 per year. This growth gives some idea of the brisk pace of research in this area.

Over the past decade with the advent of genomic sequencing, a second phase in genetic disease discovery has dawned, involving the study of susceptibility genes for a wide range of medical conditions ("multigenic diseases"). In this case no single mutant gene is causing the disease, but rather a whole range of genetic variants that, taken together, contribute to the likelihood, or not, of a particular disease developing.

In terms of research, single-gene diseases are like the low-hanging fruit that can be recognized and picked relatively easily. A large enough family history can soon reveal whether a disease is inherited, although uncovering the culprit genetic mutation can take a bit longer. By contrast, the multigenic diseases represent fruit that is much harder to reach since each variant gene may contribute only a tiny proportion of the total risk factor.

Clearly we cannot provide here a complete overview of every type of genetic disease, so we start by giving some examples of the low-hanging fruit that illustrate some of the different kinds of mutations that cause disease. We then consider the multigenic diseases that in practice represent the really big killers in the Western world.

Diseases Caused by Single Gene Mutations

Cystic Fibrosis—a Recessive Disease

Paul Quinton was nothing if not enthusiastic once he had determined to crack the mechanism underlying his own hacking cough and salty skin. Stories are told about the way in which visitors to his ranch would have fresh sweat gland samples taken using a corkborer so that Quinton could find the reason for his own salty skin. Eventually this led to the insight that something was wrong with the transport of chloride ions across the skin, leading to the accumulation of excess sodium chloride (salt) in the sweat.[5]

The race was then on to find the mutated gene that caused the condition, this in a pre-genomic era in which the only way to crack such problems was to work with multiple families suffering from cystic fibrosis. The research teams were working blind: there was no funny-colored urine or abnormal protein accumulating in the lungs that would give some clue as to what gene might be involved. Francis Collins later described the hunt as "looking for a single burned-out light bulb in the basement of a house somewhere in the United States."[6] This was the real start of "reverse genetics," the process whereby researchers hunt for genetic markers, variant portions of the genome, that are consistently inherited along with the disease, until finally the location of the rogue gene is narrowed down.

By 1985 several groups had traced the gene to Chromosome 7, but Francis Collins' group finally cracked the problem and identified the defective gene that was subsequently sequenced in the late 1980s.

The culprit turned out to be a mutant form of a gene that encodes a protein called the cystic fibrosis transmembrane conductance regulator (CFTR). In the great majority of cases, it was found that three base-pairs were eliminated in the mutant version, removing an amino acid at the 508th amino acid position in the protein. This tiny change in turn means that the protein is not folded up properly and loses its usual function. This mutation alone accounts for 90 percent of the cases found in the United States. However, more than sixteen hundred different mutations have now been described in the CFTR gene, producing various degrees of abnormality in the resulting protein. This helps explain why sufferers such as Paul Quinton could live so much longer than others with the disease. It all depends on what mutation is present. In fact, in Quinton's case only one of his CFTR genes carries the common 508th amino acid mutation, whereas his other copy has a mutation that causes milder effects. Just enough functional CFTR is present to make a positive difference.

Even in people with the 508th mutation present on both copies of the CFTR gene, the disease varies considerably in severity, with some sufferers dying at sixteen and others having quite healthy lungs well into their twenties. The reason for this variation is the action of modifier genes that impact the functioning of the CFTR gene product. Cystic fibrosis, like many other "single-mutant-gene" diseases, is not really a single gene disease after all; rather a whole array of other genes are likely to be involved in the level of severity of the disease, even though there is still one main culprit.

Progress in the general medical treatment of the disease has moved much faster than has the use of the new genetic information to introduce effective gene therapy, increasing the life expectancy to an average fifty years for those born with the condition in the twenty-first century. The vast majority of those affected now graduate from high school and can enjoy vastly improved lives compared to sufferers of only a couple of decades ago, albeit lives that can involve taking up to fifty pills a day to keep the symptoms under control.

Ironically the identification, many years ago now, of the mutant CFTR gene has not yet led to complete agreement about the way its defective protein product results in the clinical symptoms, which include the accumulation of a thick mucus in the lungs and associated bacterial infections. The CFTR protein is responsible for transporting chloride ions out of the sweat glands into the surrounding fluid, which is why the sweat is so salty when this process is blocked. But exactly how this leads on to the other range of symptoms remains to be clarified. The identification of the gene product is a major step forward in understanding, but it is only one step along the way, and has yet to lead to effective gene therapies. As Jack Riordan, a scientist who collaborated in the original discovery of the defective gene, puts it, "The disease has contributed much more to science than science has contributed to the disease."[7]

Huntington's—a Dominant Disease

There are many examples of single-gene diseases that display a dominant sex-independent type of inheritance. We encountered the classic example of familial hypercholesterolemia (FH) in chapter 1. As explained there, a dominant pattern of inheritance is observed when a single mutant copy of a gene causes clinically measureable effects. The single normal copy may be insufficient for proper functioning, or perhaps the single mutant copy interferes with proper functioning.

First described by the American physician George Huntington in 1872, Huntington's disease—a syndrome that typically starts developing between the ages of thirty-five and forty-four—provides a striking example.[8] The first symptom to be noticed is known as "chorea," the loss of normal muscular control so that movements become jerky and uncoordinated. Death usually occurs around twenty years after the first onset of symptoms, although this is not directly due to the genetic mutation but to its secondary causes: accidents due to falling, heart disease, and pneumonia. Here is a sword of Damocles hanging over someone who is clinically normal for much of one's life, but who then faces years or decades of inevitable decline, with some treatment for the symptoms available, but with no cure on the horizon at present.

The disease affects about five to ten individuals out of every one hundred thousand in the U.S. and European populations, but only one out of every million of those of African or Asian descent. There are also pockets of very high incidence in different parts of the world. For example, in the isolated populations of the Lake Maracaibo region of Venezuela, the disease affects up to seven thousand per million people. Such local concentrations of disease are often caused by "founder effects." Isolated populations may be established by a small handful of families, and if the mutation is present in the founders, then it will spread through succeeding generations, especially if there is little outbreeding.

What is going on at the genetic level is actually very strange. In 1991 a new type of mutation was discovered that causes disease. This consists of repeats of three base-pairs in a row. Normal individuals typically have fewer than thirty of these repeats on any one chromosome, but those with clinical symptoms have from thirty-five up to several thousand. At least twenty-two different diseases have now been described that are caused by these out-of-control repeats, some inherited in a recessive mode and others in a dominant mode of inheritance.[9]

Huntington's disease is one of the examples displaying dominant inheritance. It is caused by multiple CAG (cytosine-adenine-guanine) nucleotide repeats (CAGCAGCAGCAG . . .) found in the first exon of the huntingtin gene which is located on Chromosome 4. The mutant huntingtin protein in the brain causes the disease. The normal CAG range is twenty-eight repeats or fewer and will not result in disease. However, in the intermediate range of twenty-eight to thirty-five repeats, there is increasing instability of the DNA during duplication, and forty-one repeats or above means that the individual will be clinically normal until their forties, during which period the disease will gradually develop. The range of thirty-six to forty repeats marks a "danger zone" in which the disease may emerge with a slower progression and less marked symptoms than in the full-blown disease. This is known as "reduced penetrance." Making precise predictions based on this number of repeats is difficult, for as with all genetic diseases other genes also play minor roles in determining age of onset and eventual severity. With very high numbers of repeats (more than sixty), the disease becomes ever more severe, and will even appear in people younger than twenty years old.

A particular feature of this group of diseases is that once the number of triplet repeats has reached a critical level, then the number continues to grow ever greater with the passing of the generations. A parent on the borderline without the disease may have a child with one extra CAG repeat that tips the child into the danger zone, so that they then develop the disease quite unexpectedly. Before the triplet repeat mechanism was understood, this was known as "genetic anticipation": it almost seemed as if the disease was being anticipated in previous generations.

Why are these strange aberrations happening at the DNA level? Once the number of repeats reaches a certain level, then the DNA in that region becomes twisted out of its normal double-helical structure to form all kinds of hairpins, triplex, and even quadruplex structures, and DNA replication can no longer occur normally. At the same time the cell throws all its many DNA repair kits into action in a desperate attempt to fix the damage. This has been shown experimentally by systematically deleting specific components of these repair systems in yeast, bacteria, flies, mice, and cell lines, and in each case the net result is marked changes in the stability of these CAG regions.[10]

The number of repeats can increase very early in development, soon after fertilization. In some triplet repeat diseases, individuals contain two sets of cells, each with a different number of repeats. This shows that after the zygote had divided once, then different numbers of triplet units were added in each of the two daughter cells, which in turn multiplied to generate the body. This explains why supposedly identical (monozygotic) twins can sometimes differ in their triplet repeat number.

Since CAG encodes for the amino acid glutamine, it can readily be seen how repeated CAG sequences in an exon could subvert normal protein function since the huge tracts of glutamine inhibit proper folding of the protein. In fact, a characteristic feature of the disease is the accumulation of polyglutamine-rich protein aggregates. Proteins that don't fold properly are a common cause of breakdown of the normal functioning of cells.[11]

But working out exactly what the huntingtin protein does in the brain has not been easy. It interacts with many proteins and many signaling pathways, making it difficult to sort out which are really critical. What is clear, however, is that the mutant huntingtin protein is toxic to cells, causing the gradual loss of key neurons (brain cells) in parts of the brain that coordinate muscular control. This gradual loss accounts for the slow onset of the disease: only when neuronal loss reaches a certain critical level do symptoms appear, and these include psychiatric symptoms as many parts of the brain become affected by neuronal loss. This feature explains why Huntington's is a dominant disease: it is not due to the lack of a protein (the result of a "loss-of-function" mutation), but rather to the gain of a new (and unwanted) function.

Huntington's disease is one of those syndromes that remain a stark challenge to the medical research community and to those families in which the mutant condition is known to be transmitted across the generations: Is it better to know or not to know that you carry the disastrous number of CAG repeats? Will gene therapy ever be possible? And should prenatal diagnosis be practiced for a disease that develops relatively late in life? These are not easy questions, and we return to some of them in the final chapter.

Chromosomal Diseases and MicroRNA Regulation

Chapter 4 surveyed the huge range of different types of genetic variation. Any and all types of variation can and do cause disease. We have already mentioned the deletion of 740,000 base-pairs of a specific region on human Chromosome 16 that leads to a heritable form of obesity. Other structural variations in chromosomes may lead to disease, although in many cases the variations do not occur in the germ line; if disease occurs, it is not passed on.

One structural variation appears to mediate at least some of its effects via microRNAs, thereby illustrating at one stroke the role of two different types of genetic control in disease. This is DiGeorge syndrome, named as usual after the physician who first described it in 1968, Angelo DiGeorge. Affecting around one out of every four thousand live births, the disease is caused by a deletion of about 3 million base-pairs from Chromosome 22. That sounds like a lot, but it should be remembered that even though this chromosome is the second smallest of them all, it still contains around 49 million base-pairs of DNA, so the deletion represents only 6 percent of the whole. But even that 6 percent contains more than twenty-five genes, explaining why the syndrome is so varied and complex; indeed, it was previously classified as several different diseases until it was realized that the Chromosome 22 deletion was common to them all.

About 90 to 95 percent of all cases of DiGeorge syndrome occur de novo, so are not inherited. The abnormality occurs in very early development, but not so early that it's present in the germ-line cells. But if the deletion does occur in either the sperm or the eggs, then it becomes an inherited abnormality and is passed on in a dominant fashion as each child receives one copy of Chromosome 22 that is missing the deleted segment. As mentioned in chapter 4, structural variations in the genome are not strictly random, occurring more often in some chromosomes than others, and the structure of Chromosome 22 at the spot where the DiGeorge deletion occurs is indeed particularly prone to such events.

DiGeorge syndrome is marked by a wide range of clinical symptoms, any one of which may be the main feature in a given patient, so chromosomal analysis is important for making a final diagnosis. Typically cases may present with congenital heart defects at birth, defects in the palate, learning disabilities, and recurrent infections due to deficiencies in immune system development. A fascinating albeit unfortunate twist in the story is an increased risk of schizophrenia in these patients by as much as twenty- to thirty-fold. Turning the numbers around the other way, if you take a population of schizophrenics, then the Chromosome 22 deletion is found in about 1 percent of cases, compared to 0.025 percent of individuals in a random population.

Unlike cystic fibrosis or familial hypercholesterolemia, where one main defective gene is involved, sorting out how deficiencies in more than twenty-five gene products cause disease is quite a tall order. Remember, too, that the products of all these genes are still present at roughly 50 percent reduced levels, so in many cases that 50 percent will probably be quite sufficient for normal function, whereas for others even a 50 percent reduction causes dysfunction. One of the genes involved in DiGeorge syndrome,[12] if disrupted in mice, causes cognitive impairment and a loss of a range of microRNAs because it plays a key role in generating these key molecules from their mature precursors. Given the important role of microRNAs in regulation (see chapter 2) it seems reasonable to suppose that a generalized microRNA deficit in the brain would contribute to at least some of the clinical symptoms in the patients. Once again this provides a nice example of the way biological systems operate: mess with one bit of the system, and all kinds of secondary effects then occur.

Susceptibility Genes and Multigenic Diseases

Although the existence of more than six thousand single-gene diseases sounds like a lot, and clearly all these diseases are often devastating in their implications for those who suffer from them, on the other hand they represent only a tiny percentage of all human disease. The really big diseases that affect large portions of the population include heart disease, stroke, cancer, diabetes, arthritis, asthma, Alzheimer's disease, and many others. With rare exceptions, none of these diseases follows any of the relatively simple recessive, dominant, or X-linked modes of genetic transmission that we have discussed so far. And yet everyone knows from their own family history, or that of their friends, the tendency exists for some types of disease to run more in one family than another.

The hunt for susceptibility genes then comes into the picture, a research field that has received a huge boost from the advent of genomics. For decades there were attempts to track down variant genes that contribute a small portion of the risk of developing a disease, but the tools available were too blunt to get very far. All that has now changed with the ability to screen the whole genome for thousands of genetic markers and then to see which particular markers are found in those family members who suffer from a particular disease, but less so in people without disease. By studying genomes from enough people who have the same disease, plus enough disease-free individuals, the small portion of the genome that contains the risk factor gene(s) can then be identified. This has led to a "genetic gold rush" in which researchers have been in a race to use the new technology to reveal susceptibility genes. Not surprisingly, biotech companies are strongly interested in the race, hunting for the next drug targets that might help to control the world's biggest killer diseases. That each investigation costs at least $1 million and that several different groups have analyzed the same disease gives some idea of the enthusiasm for this new strategy.

The approach is known as Genome Wide Association (GWA), and the genetic markers are the Single Nucleotide Polymorphisms (SNPs) that act as signposts scattered across the genome. Hundreds of thousands or even several million of these are used as flags to mark different segments of the genome that vary slightly between individuals. The GWA approach is based on the assumption that common diseases are attributable in part to variant genes that are present in at least 1 to 5 percent of the population. Some of the marker SNPs are located within protein-coding genes, but most are not, instead acting as signposts to regions of the genome that need further investigation.

The use of GWA to investigate the genetic causes of variation in height gives a good example of some of the challenges involved, and certainly height does vary quite dramatically, as Figure 9.1 illustrates. Height is a good place to start as it's simple to measure and adult offspring are, on average, roughly of a height intermediate between that of their parents, providing a constant visual reminder of the inheritability of this trait. In fact, genes contribute as much as 80 percent to variation in stature.[13] Several GWA height studies have been carried out, one on more than 180,000 individuals revealing no less than 180 different variant SNPs that flag small regions of the genome contributing to height variation. Yet even when the effects of these 180 different regions are added together, they only account for about 10 percent of height variation.[14] There is, therefore, a huge amount of "missing heritability." Where is it?

One problem is that even with a very large number of SNPs in the study, the whole genome is a very large space to cover completely with marker flags. A given SNP may still be a long way from the gene that is making a difference in height—so far indeed that after a few generations it may get separated from the critical gene that's making the difference. Don't forget the random exchange of DNA segments that takes place between paired chromosomes dur-

FIGURE 9.1. Both genetic variation and phenotypic variation in height and other quantitative traits tend to be greater in populations of (recent) African descent than in populations of European or Asian descent. This is illustrated by the greatest height difference on record between teammates in the National Basketball Association. Manute Bol (7 feet, 7 inches) and Muggsy Bogues (5 feet, 3 inches) both played for the Washington Bullets in 1987. Reprinted by permission from Getty Images. Photographer: Jerry Wachter

ing meiosis ("recombination"). Furthermore, even when a relevant gene has been identified, it may be a transcription factor that in turn regulates the expression of lots of other genes located far away on the genomic landscape, so plenty of sleuthing is necessary to find out what's really happening.

A further explanation for the missing heritability is the fact that height variation might look simple because we're so used to it, but hundreds, perhaps thousands, of different genes are involved in its regulation. Think about being very tall: bones have to be longer, muscles longer, feet a bit bigger, all the internal organs stretched a bit further, the heart pump adjusted to cope with pumping blood a bit higher, and so forth. A huge coordinated operation is necessary to adjust for height, and each gene might contribute only a tiny percentage of the total 80 percent heritability.

In the case of disease, heritability is calculated by simply tracking how often a disease reappears in different family members over many generations in comparison with its random appearance in any population. As with the case of height, GWA studies on disease generally tend to reveal far less than 50 percent of the genetic variation that explains the known heritability of the disease based on family studies, and often less than 10 percent.[15] One reason may be rare genetic variants, such as small deletions or insertions, that explain much of the disease heritability in a given family history but are missed by GWA studies.

The challenge of tracking down heritability is well illustrated by the results from a huge study involving more than one hundred thousand individuals with the aim of identifying the variant genes that account for differences in blood lipid levels. As noted in chapter 1, lipid regulation plays a key role in cardiovascular disease. The study revealed no fewer than ninety-five gene variants, yet these still explained only about one-quarter of all the heritability known to exist for blood lipid levels.[16] It is unlikely that even larger cohorts will find the missing 75 percent heritability in this case; scientists are hunting for the rare variants of large effect, or small DNA deletions or insertions, that might help solve the mystery.

Other GWA studies have been luckier in terms of quickly identifying key genetic variants that cause disease. For example, one investigation was carried out to find the genes that regulate either very low or very high fetal hemoglobin.[17] Remember that hemoglobin is the protein that carries oxygen around the blood and delivers it from the lungs to all the other tissues of the body. A special version of hemoglobin is used in the fetus, and its production is usually switched off soon after birth as the adult version kicks in. However, some people retain rather high levels of the fetal version into adulthood, and it does them no harm. In fact, in patients suffering from blood disorders such as sickle cell disease or thalassemia, high fetal hemoglobin can actually protect them against some of the diseases' worst symptoms, such as leg ulcers, pain, and even death. Hence the biomedical research interest.

This particular GWA study illustrates well the power of the approach since SNPs linked to high levels of fetal hemoglobin are located within a quite unexpected gene with a boring name: BCL11A, located on Chromosome 2, encoding a transcription factor. But the role of BCL11A is anything but boring. Up until the time of this research, BCL11A had been known to be involved in the progression of certain cancers, and its involvement in hemoglobin regulation was only revealed by the GWA study. Follow-up biological investigations revealed that normal BCL11A is indeed involved in switching off fetal hemoglobin production during development, so it acts as a repressor of gene expression. This finding in turn has led to pharmaceutical interest to see whether any drugs could switch off BCL11A artificially, thus repressing the repressor so that fetal hemoglobin levels could provide their valuable protective effects in patients with blood diseases such as thalassemia. This account provides a good example of the way in which genomic studies can quickly identify key genes, leading in turn to investigation of their function, so providing a rational basis for drug discovery.

Unfortunately, however, GWA studies carried out on some of the biggest diseases that afflict humanity have not generally yielded such immediate insights into the disease process. At the time of writing, more than 450 GWA studies have been published, and the associations of more than twenty-six hundred SNPs with different syndromes or diseases have been reported.[18] The numbers will continue to explode, driven by ever speedier and cheaper technologies. Many studies are carried out by large collaborative groups of research scientists from dozens of different countries, such as the Wellcome Trust Case Control Consortium and the Breast Cancer Association Consortium. The collaborations have speeded up the work and provided valuable cost savings.

Taken overall, GWA studies have moved the investigation of the genetics of disease to a whole new level. At the same time, the studies have thrown up enough research challenges to investigate the functions of novel genes for many years to come. The results continue to tease our desire to gaze into the crystal ball and predict our futures. Commercial DNA tests are based on analysis of the most common SNPs associated with risks for different diseases. The problem is that the GWA studies are averages based on large sample size and do not necessarily make any kind of accurate prediction for a specific individual. At best, such tests can provide a generalized percentage lifetime estimate of risk for developing a given disease. Given that usually we cannot do much about it anyway, do we really want to know? This latter question becomes even more pressing when we are considering the single-gene mutations that certainly cause disease, such as Huntington's disease, but which don't appear until later in life. Genetic knowledge is rarely neutral in its implications.

Cancer

"The knowledge that I have cancer never goes away. I am aware that I am dying every moment of every day, and every waking moment of every night. Each day is new territory for me: will I get up in the morning and find I cannot control my bowels? . . . The trial is knowing that I will never feel 'better' again, and that edge is going from all physical pleasures." These words come from a friend dying from an inoperable case of gallbladder cancer. He died a few months after writing these words.

The word "cancer" brings an understandable dread to every human heart. We all have friends or relatives who have died from this disease. One out of three of us will suffer from cancer sometime during our lifetime. Every year 462 out of every 100,000 American men and women are diagnosed with cancer.[19] The five-year survival rate is currently running at 67 percent on average, but this average figure is not very useful as survival rates vary hugely depending on the type of cancer, age, time of diagnosis, and many other factors. In reality, cancer is not one disease but many, yet they all have in common that they are diseases of the genes.

More than two hundred different types of cell make up the tissues of our bodies. Virtually any type of cell can start proliferating out of control; when this happens, it is called a cancer. The prognosis for cancer depends very much on what particular cell type is affected and whether the cancer has spread to other tissues, known as metastasis.

Why does a cell start proliferating out of control? In every case it involves DNA damage of some kind. It is estimated that each gene is mutated once in every twenty thousand cells.[20] The vast majority of these cells are somatic cells, so this DNA damage is generally localized rather than inherited, but once damage occurs there is always the potential for more damage and a cascade effect. A normal cell is swift to recognize the DNA damage and then usually switches into a "pause" mode, during which further replication is on hold, so allowing time for DNA repair before any further replication takes place. However, if DNA damage is sufficiently bad, then a different program is switched on known as "apoptosis," or programmed cell death, in which the cell commits suicide rather than carrying on with a damaged genome.

The problem is that sometimes apoptosis is prevented by "oncogenes," genes that have mutated such that they are no longer under proper regulation or no longer regulate other genes in a normal way. If apoptosis is prevented, then the damaged cell carries on replicating and more damage accumulates, until finally the cell becomes locked into a state of continuous replication, giving rise to a full-blown cancer. Such cells then begin to dominate by a process akin to Darwinian natural selection in which they keep on dividing, swamping the normal cells.

We encountered oncogenes in chapter 4, in the form of fusion proteins that are generated by the swapping of segments of DNA between two different chromosomes. One of these fusion proteins is called BCR-ABL, the ABL segment of DNA arriving from Chromosome 9 to find itself next to the BCR segment on Chromosome 22, there effectively forming a new BCR-ABL gene, which in turn causes chronic myelogenous leukemia, a cancer of the white blood cells. The problem is that the BCR portion hyperactivates an enzyme called a kinase that is encoded by the ABL portion, which drives the cells into the cancerous state in all kinds of ways. One action of BCR-ABL is to inhibit apoptosis so the cells with damaged DNA stay alive and keep dividing.

In this particular cancer, the cells become addicted to the presence of BCR-ABL and cannot keep proliferating in their cancerous state without it. That part of the story gives hope, because a drug called Imatinib (Gleevec) selectively inhibits the actions of the BCR-ABL kinase. This drug and its even more powerful successors have done much to transform the outcome for patients suffering from this disease. It is a genetic disease, but "genetic" in quite a different kind of way from the diseases discussed in the previous section.

Some gene mutations are inherited in germ-line cells, giving rise to a significant degree of inheritability in certain cancers. Indeed their effects are so striking that such genes were discovered long before GWA studies came along. An example is provided by the BRCA-1 and BRCA-2 genes ("breast cancer early onset genes"). Women with mutations in these genes have about a 60 percent risk of developing breast cancer before the age of ninety compared to 12 percent in a population with nonmutated BRCA genes. The risk of ovarian cancer is also greatly increased: the lifetime risk for women carrying the BRCA-1 and BRCA-2 mutations is about 55 percent and 25 percent, respectively, compared with 1.8 percent in the nonmutated population. Despite these striking statistics, don't think that most breast cancer comes from BRCA gene abnormalities: far from it. BRCA mutations are rare, occurring in only one out of one thousand people, and accounting for less than 5 to 10 percent of all breast cancer cases, and about 10 to 15 percent of ovarian cancers.

BRCA-2 is one of the few genes that you can ride over on your bicycle (perhaps unique in this respect). Inspired by the fact that the gene was sequenced at the Sanger Institute near Cambridge (United Kingdom), a bicycle path has been built from the Addenbrooke's Hospital research campus, a few hundred yards from where this is being written, to the nearby village of Little Shelford, where many biomedical researchers have their homes. Embedded in the path is the complete BRCA-2 gene sequence in colored tiles representing the four nucleotide bases, 10,257 nucleotides altogether, enough for a very good bike ride. The best time to ride the path is on a moonlit summer night as the path through fields is lined by solar-powered battery lights that glow to mark the way (preferably a wind-free night, as a basic law of physics is that the wind is invariably against you when cycling). A sign by the path reminds you that on this scale the whole genome would circle the earth fifteen times and it would take a cyclist three to four years without stopping to traverse it—a sobering thought.

The proteins encoded by the BRCA genes are involved in DNA repair, among other functions, and so are known as "suppressor genes" because they prevent too much DNA damage from accumulating. But the mutant forms of BRCA are unable to carry on this function properly, allowing DNA damage to accumulate and causing cancer. More than a thousand different mutant forms of BRCA genes have been described.

As with other diseases, GWA studies have also made a significant impact on the understanding of much smaller variant gene contributions to different cancers. Examples of single-gene mutations that greatly increase cancer risk, like the BRCA genes, are relatively rare in cancer studies. Much more common are those cancers influenced by the effects of multiple-susceptibility genes. For example, whole series of new SNPs have been associated with prostate, colorectal, lung, breast, skin, and brain cancers, among others.[21] Occasionally such studies reveal one variant that is a risk factor for one cancer but which actually helps to suppress a quite different disease. For example, the presence of the variant TCF2 increases the risk of prostate cancer, but at the same time decreases the risk of type 2 diabetes, a type of diabetes in which cells become resistant to the actions of insulin.

The advent of cheaper genomic sequencing has also spurred major efforts in the sequencing of tumor DNA from hundreds of different patients with clinically identical cancers in an attempt to identify the real drivers of the disease. The results have revealed a vast, bewildering array of mutations in different tumors. As mentioned above, once DNA damage gets a grip on a genome, then it tends to spiral out of control. Tumors consist of cancer cells that have gone through multiple rounds of replication; with DNA repair mechanisms subverted, the genome becomes increasingly damaged.

In one study carried out on a breast cancer tumor in which the complete tumor genome was sequenced from a forty-four-year-old African American patient, 27,173 point mutations were discovered in the primary breast tumor, of which only 200 were in actual gene regions.[22] An interesting aspect of this study (though tragic for the patient) is that the tumor metastasized to the brain after a period of eight months. A sample from this tumor was sequenced also and found to contain 51,710 mutations, nearly twice as many as in the original tumor. The investigators then also grew some of the breast tumor inside a mouse, and subsequently sequenced the tumor that grew there, now finding no fewer than 109,078 mutations in its genome, four-fold more than in the original tumor. Most mutations were common to all three tumors. One of the informative insights that came out of the analysis of this great flood of data was that in each tumor examined, a gene that plays a role in suppressing metastasis was mutated, helping to explain the readiness of this cancer to spread to the brain. Those who die from cancer and are willing for their bodies to be used for medical research may be helping thousands of people in the future to benefit from their sacrifice.

The big challenge is to find the real drivers of cancer among all those thousands of irrelevant mutations that really make no difference to the continued cell replication and survival that characterize the cancer. Only by screening thousands of tumors and many thousands of mutant genes is it going to be possible to find the needles in the haystacks. But the hunt is on, the international resources are huge, and eventually the great outpouring of data will provide new biological insights into tackling the world's most feared group of diseases.

Gene Therapy

When molecular genetics really began to take off in the 1960s and 1970s, we all thought that gene therapy was just around the corner. Obviously if there were mutant genes, then it shouldn't be that difficult to patch them up or replace them. If transferring genes into cells in the laboratory was so simple, which it is, why couldn't that be the case also for replacing defective genes in human tissues?

By the early 1990s the first serious gene therapy trials were under way, but all the experience since has shown that gene therapy is far more challenging than anyone had originally imagined. Having said that, recent years have seen the rather gloomy atmosphere gradually change to one of hope and expectation, mainly due to some promising trials. Successful gene therapy is certainly not yet routine for any disease, but there is some room for cautious optimism.

Why is it so difficult? For a start, it should be emphasized that the aim of all trials so far has been to provide somatic tissues with a good supply of the missing or defective gene. Any attempt to change the inheritable genome in germ-line cells is currently illegal in the United States, as in many other countries.

To get a missing gene into somatic cells one needs a good vector—that is, a carrier that will contain the gene, gain entry into the cell, and then incorporate the gene with reasonable efficiency into the patient's cells' nuclear DNA. But that is only the start of the challenge. If the therapeutic gene is incorporated randomly into the genome, it might end up in a place where it is not expressed properly, or will disrupt other genes or even cause cancer. Furthermore, the number of copies of the gene incorporated may be too high. As if all that wasn't enough, the body's immune defense system may recognize as foreign the vector used to transfer the gene into the diseased cells and attack the vector viciously.

None of these challenges is at all theoretical. They can all be illustrated by one or more of the various clinical gene trials carried out so far, and hundreds more trials are in progress. The experience with cystic fibrosis provides an example. As soon as the mutant gene was identified, clinical trials were in mind, although they were not launched until 1993. An early challenge that researchers faced was that the lung is highly specialized to identify and reject foreign bodies, as you might expect from an organ that breathes in more than ten thousand liters of outside air every day. Then in 1999 a severe immunological reaction killed Jesse Gelsinger in a gene-therapy trial in the United States for an inherited liver disease, casting a shadow over the whole gene-therapy field for many years.

The hopeful signs began to come from an unexpected direction.[23] Rather than replace the defective gene directly, went the reasoning, why not use a therapy that helps the mutant protein to function properly? Even if your car is not functioning properly, you could still hire a truck to give you some assistance. As mentioned in the previous chapter, the main problem with the CFTR protein is that it doesn't adopt its correct shape if certain of its amino acids are changed due to a mutation in the CFTR gene. A "corrector protein" (the "truck" in the analogy) can help resolve the problem by helping the protein to fold properly. In some mutants the problem is slightly different in that the protein folds up correctly, but when it reaches the cell membrane, it doesn't function unless helped along by another protein (a second "truck"), known as a "potentiator." In 2008 it was announced that the gene for a potentiator known as VX-770 had been introduced into the lung cells of cystic fibrosis patients with a rather rare CFTR mutation, producing some striking clinical benefits. As Robert Beall, director of the Cystic Fibrosis Foundation in Bethesda, Maryland, reported at the time, "It was the most emotional time since the discovery of that gene. It's telling you we can change the course of this disease."[24] "Corrector" protein genes are also undergoing clinical trials, so it could be that a double whammy of both corrector and potentiator proteins just might do the trick. If your car is in really bad shape, then one truck pulling and another pushing might be the answer.

Meanwhile, attempts are still being made to deliver the CFTR gene directly to the lung cells that need its protein product so badly. One delivery vehicle is known as "liposomes." These little fat (lipid) vesicles less than a micron across can be used to transport the "healthy" CFTR gene. Liposomes were originally discovered by a biochemist named Alec Bangham at The Babraham Institute in Cambridge (United Kingdom).[25] At the time they were not thought to have any commercial use, but since that time have found a wide range of applications, not least in drug delivery. Their advantage is that they fuse with the lipid membranes of cells and deliver their contents to the cell interior. Liposomes provide a good example of how basic research can lead to unexpected applications, sometimes many years after the initial discovery.

The stakes are high with gene therapy. Cure the disease at the source, and then there's no need to worry about the symptoms. But meanwhile the treatment of the symptoms is helping cystic fibrosis sufferers live ever longer and healthier lives. And sometimes a little serendipity can help with that, too. Some years ago doctors in Australia realized that their patients who surfed did better (on average) than those that didn't—due, it emerged, to the daily inhalation of super-salty water, which has now become a standard part of the treatment regime.

Some of the biggest successes in gene therapy trials have come with patients suffering from rare immunodeficiency diseases, meaning that they have virtually no defense against pathogenic viruses and bacteria. These have become well-known because of the vivid pictures of "bubble children" totally enclosed within a transparent tent that provides them with a pathogen-free environment (Figure 9.2). For some, at least, gene therapy is providing new hope with a normal life back at home and no need for such isolation. But it has been a long and winding road to make it this far.[26]

FIGURE 9.2. "Bubble Child" David Vetter. David suffered from severe combined immunodeficiency disease (SCID) and lived for twelve years inside plastic "bubbles" that protected him from any infectious agent. Today, children born with SCID may be treated with one of the most effective gene therapies known to date. Reprinted by permission from Baylor College of Medicine Archives.

The best-studied condition is known as severe combined immunodeficiency disease (SCID). This disease results either from a mutation in the gene encoding the enzyme adenosine deaminase (ADA) or, in its X-linked inherited form, from a defect in the gene encoding a key structural protein component of different receptors that mediate the action of cytokines, important regulatory messengers of the immune system. It has been known since the 1980s that transfer of the stem cells found in the cord blood of a tissue-matched "savior sibling" (so-called) can be therapeutically effective in such cases. Relatively few transplanted stem cells can expand to effectively reconstitute a whole new immune system. Nevertheless, in the great majority of cases, tissue-matched donors are unavailable, and the danger of rejection of the transplanted cells still is present, even when they are closely matched to the patient's own tissues. But if the normal ADA gene could be added to the sufferer's own stem cells (obtained from a sample of their bone marrow) and transferred back into the body, then there would be no problem of rejection.

Several successful trials using precisely this approach have been carried out since 2000, first in Paris and Milan, then in the United Kingdom and United States.[27] For SCID due to adenosine deaminase deficiency, some excellent results have been obtained: around 80 percent of the children treated for the disease have seen long-term remission and been able to return home to live normal lives. For X-linked SCID, the situation has been more complicated: although eighteen of the twenty children who originally underwent gene therapy have thoroughly reconstituted and functional immune systems, a leukemia-like syndrome developed in five of the children within a few years of therapy. Sadly one of the children died. The leukemia seems to be caused by the therapeutic gene inserting into the genome in such a way that it switches on one or more nearby oncogenes.

The experience with SCID therapy illustrates the challenges, setbacks, and promise of gene therapy. The effects of any gene are always dependent upon context, being mediated through a complex network of interactions. But as therapeutic experience accumulates, based on hundreds of trials, so the positive impact of gene therapy will increasingly be felt in medicine in the decades ahead. But it may yet prove a difficult road.

CHAPTER 10

Guarding Our Genomes

The Impact of Epigenetics

EXPLORING THE FAR NORTH of Sweden might not be the first project that comes to mind when investigating the fascinating world of epigenetics. In a windswept county called Norrbotten just outside the Arctic Circle lies nearly a quarter of Sweden's land area, but with only an average of six people living in every square mile (compared to more than twenty-seven thousand per square mile in New York City). So isolated was it in the nineteenth century that if the harvest was bad, people starved. But in the years of good harvest, people would eat well for months. From a research perspective, the most useful part of the story comes from the extensive historical data on harvests and food prices, demanded by His Majesty the King of Sweden, that go back as far as 1799, together with the parish records that exist to the present day, recounting not only the years of good and bad harvests but also the births and deaths of the inhabitants.

In the 1980s a health care specialist named Lars Olov Bygren wondered what might have happened to the health of children growing up during these times of famine and feast, and not just to them but also to their children and grandchildren. Teaming up with a London geneticist named Marcus Pembrey, the results that emerged are fascinating and have continued to pique the interest of the scientific community and beyond ever since. The researchers followed randomly selected cohorts of children who were born in the parish of Överkalix in 1890, 1905, and 1920 and whose two parents and four grandparents could be identified. They found that grandsons lived shorter lives if their paternal grandfathers experienced abundant harvests in their youth.[1] In fact, they died on average much earlier than grandsons whose paternal grandfathers had been raised in times of famine. Longevity trends were also seen in granddaughters whose paternal grandmothers had been exposed to poor nutrition versus years of plenty.

A lower incidence of cardiovascular disease seems to have been a significant factor in those who lived longer, and a higher incidence of diabetes in those who lived shorter lives.[2] In fact, the incidence of diabetes was four-fold higher in those whose grandfathers had experienced abundant eating. The results could not be explained by "longevity genes" in one set of grandfathers compared to another because, irrespective of their diet, the average age of the grandfathers when they died was remarkably similar.

Somehow the genes carry a "memory" of an environmental influence that occurred two generations earlier. How come? One important clue comes from the precise ages of the grandparents during the times of feast or famine. The "slow growth period" was estimated for that historical period as being nine to twelve years old for boys. This is the time when children's development slows down, just before the growth spurt associated with puberty. The paternal effect on the grandson's longevity was only observed if the grandfather, when a child, had been exposed to feast or famine during his critical "slow growth period," and not at other times. Why is that critical? Because during this period sperm development is in full swing in preparation for puberty.[3] But when the grandmotherly effect down the maternal line was investigated, it was found that the biggest impact on the health of the granddaughters was when the grandmother had been exposed to feast or famine at the age of zero to three.[4] Egg development in the human female takes place largely during fetal development and soon after birth, so again the "memory" of feasting or fasting seems to be related to the time of gonadal development. The maternal and paternal patterns of inheritance also point to a possible role for the X and Y chromosomes, respectively.

Another sex-specific environmental effect of quite a different kind, "transmitted" from fathers to sons, has also been noted in the Avon Longitudinal Study of Parents and Children.[5] In this study, carried out in the Bristol area of western England, the outcomes of 14,024 pregnancies were investigated in great detail. A striking finding is that for those fathers who had started smoking early, before eleven years of age, their boys, but not their girls, had a significantly higher body mass index at the age of nine. In other words, they were fatter, in itself increasing the likelihood that they would be more obese in adult life. The boys were progressively less fat with the increasing age at which their fathers had started smoking. Once again an environmental influence during the slow growth period appears to have made an impact later, in this case in the next generation. Everyone knows that smoking is bad for you. But it might not occur to many boys who start smoking early that their habit might just be condemning their future sons, as yet unborn, to increased risks of ill health in adulthood.

Further fascinating biological insights are provided by studies carried out on those who were in the wombs of Dutch women during the German occupaation of the Netherlands in 1944–1945, known as the "Hunger Winter" in the Netherlands.[6, 7] In neither this or the other examples already mentioned is it clear at a molecular level how an environmental change in one generation affects the health and well-being of descendants two generations later. Indeed, cross-generational social influences are so powerful in shaping human behaviors and human health that it is always difficult to exclude them. Nevertheless, the publicity surrounding such studies perhaps more than anything has drawn attention to the burgeoning field of epigenetics, and epigenetic mechanisms remain the favored explanation, for without doubt they explain comparable changes that are now known to occur across generations in both plants and animals.

What Is Epigenetics?

You can always tell that it's a new field in science when the authors of scientific papers routinely define the meaning of their field in the introduction. This remains the case with the field of epigenetics, and the various uses of this somewhat slippery term can be confusing for the nonspecialist. Plenty of examples are provided below to see what epigenetics means in practice.

The term "epigenetics" comes from combining the Greek *epi*, meaning "over" or "above," with "genetics." Biologist C. H. Waddington coined the term in 1939 to refer to "the causal interactions between genes and their products, which bring the phenotype into being."[8] Waddington's use of the term was so broad that it would include most aspects of what we now refer to as developmental biology.

Today epigenetics refers to all those inheritable changes that occur in chromosomes that bring about changes in the phenotype without any changes occurring in the DNA nucleotide sequence. "Inheritable" in this context may simply refer to the transmission of information from parent cells to daughter cells during replication, or it can refer to inheritance across generations. If genetics provides the hardware of inheritance, then epigenetics provides the software. Or if the DNA sequence is the musical score on the page, then epigenetics is all those Italian words (*fortissimo* and the like) that give instructions on how the notes should be played.

At its most basic level, epigenetics refers to the ways in which the cells of the very early embryo, with its "totipotent" cells that have the potential to develop into any cell of the body, are then gradually differentiated to form the two hundred or so tissues that make up human bodies. We take it for granted that this hugely complex process is on automatic and we don't have to think about it. If we did, most of our brain power would be needed to send out the necessary instructions, because thousands of genes have to be switched on and off in a permanent fashion in order to make a brain cell, a blood cell, or a liver cell. If skin cell instructions somehow got mixed up with blood cell or brain cell instructions, we might end up looking like squidgy tomatoes or stale porridge.

How Are Epigenetic Changes Transmitted?

At a molecular level there are three known mechanisms that mediate epigenetic changes: methylation, chromatin modification, and microRNA, and the resulting chemical changes in the DNA are known as "epigenetic markers." We have encountered all these mechanisms before, and in chapter 3 we saw how the first two of these mechanisms are intimately involved in the cellular differentiation that occurs during the early weeks of the developing embryo. As pointed out in that chapter, brain cells need about 60 percent of their total number of protein-coding genes to function properly, whereas skin cells need only about 40 percent to be switched on. What is important in this context is that the modifications of the genomes contained in each specialized cell need to be permanent. Once development into a skin cell occurs, then skin cells need to give rise to other skin cells, and not to kidney cells or brain cells. In other words, epigenetic changes need to be heritable. To see how that happens, we need to delve a little more into the mechanisms involved.

As already mentioned in chapter 3, methylation involves the transfer of a methyl group to a cytosine base. A methyl group is very simple, consisting of one atom of carbon joined to three atoms of hydrogen (CH_3). When cytosines in a gene promoter are methylated, this causes the gene to be "silenced" or "switched off." In the human genome, approximately half of gene-promoter regions contain stretches of sequence that are CpG-rich. This means that a cytosine has a guanine next to it, and the "p" refers to a phosphate group that links the two together in the long chain of nucleotides making up the DNA. Being followed by a G makes a C more likely to be recognized as a site for methylation, and gene promoters often contain lengthy clusters of CpG repeats known as "CpG islands." Conversely, removal of the methyl groups, known as demethylation, can be used to switch a gene on again. In mammals 70 to 80 percent of CpG units are methylated at any one time, whereas only 2 to 5 percent of all cytosines are methylated (remember that most cytosines don't have a G next door).

Methylcytosines are particularly prone to be mutated to thymines, as mentioned in chapter 4. Despite the fact that in human DNA only a few percent of the nucleotides are methylcytosine, about 30 percent of all point mutations are found at these sites. The epigenetic modifications that increase cytosine methylation also increase the chance of permanent DNA mutations occurring at that particular position in the DNA sequence. Once again this underlines the fact that mutations in the DNA nucleotide sequence are not wholly "random" in that they don't occur with equal probability throughout the genome. Epigenetic modifications can "channel" mutations so that they are more likely to occur in one place than another.

When a cell becomes a liver cell, a kidney cell, or whatever, without reverting back to something different, then at least part of this commitment is regulated by the methylated CpG islands, which keep whole arrays of genes permanently switched off. Of course, during our lifetime our liver cells are constantly replicating to make more liver cells to keep the organ in good shape. So how can the patterns of cytosine methylation in the DNA be faithfully passed on to the daughter cells so that they retain the same profile? Here is where the double-helical structure of DNA is once again so useful at making possible the communication of information. As we described in chapter 1, when a molecule of DNA replicates, then each strand serves as a template for the synthesis of a new daughter strand. The template strand contains the methylation pattern, and certain enzymes specialize in recognizing methylated cytosines in the template. As soon as the new CpG unit is made in the daughter strand, the enzymes simply pop a methyl group on to the newly made CpG in exactly the right spot, so that the new strand is not only the exact replica of the template at the nucleotide level but also contains methylated cytosines in the right places as well. Brilliant. But the fidelity of this replication mechanism is not nearly as accurate as for replication of the DNA double-helix itself. The error rate is about one in a thousand for the transfer of epigenetic marks through a cell division, whereas for the nucleotide sequence, the error rate is more like one in a million.[9]

We can now begin to see how the kind of environmental influences that occur during pregnancy or at puberty, as in the Swedish, Bristol, and Dutch examples mentioned above, might exert their effects, leading to epigenetic changes in the offspring that can affect their own health and even that of their progeny. For example, the increased ill-health of those whose mothers were pregnant during the Dutch Famine has been linked to an undermethylated version of the gene for insulin-like growth factor 2, a hormone (chemical messenger) involved in growth of the fetus during gestation.

The other main mechanism for epigenetic modification involves the regulation of chromatin. Recall from chapter 3 that chromatin is the intertwined mix of DNA and proteins that make up the chromosomes. Whether the chromatin is in a very tightly bound or more loosely bound "open" state determines whether transcription factors have access to genes to switch them on and off. We now need to add just a little more molecular detail to this general picture in order to see how chromatin modifications function in the transmission of epigenetic information.

One of the problems of iconic model images is that we can end up thinking that reality really looks like the model. This can readily happen with the DNA double-helix. For example, in two locations along the Cambridge bike path introduced earlier, there are two modernistic metal sculptures of the double-helix. Glinting in the sun they look so firm and durable. But of course DNA in the cell is nothing like that. It's twisted and packaged in incredible contortions along with its protein partners to fit into a tiny space. And it's not static at all but a hive of dynamic activity. Imagine a busy

DNA dockside with ships constantly moving in and out, loading and unloading their cargoes, and that's more the picture.

The DNA "protein partners" are known as histones. In cells that have a nucleus (eukaryotes),[10] as Figure 10.1 illustrates, 147 base-pairs of DNA are wrapped around eight molecules of histone that come in pairs. This little unit is known as a nucleosome. Chromatin consists of lots of nucleosomes packaged together. Each histone has a tail that sticks out from the nucleosome. That is really significant because it enables enzymes to make chemical modifications to the tail that in turn change the structure of the nucleosome. One important modification is known as "acetylation," which involves the transfer of a CH_3CO group to an amino acid called lysine in the histone tail. Note that this chemical unit is just a methyl (CH_3) with an extra atom of carbon (C) and oxygen (O), so it's really quite simple. For quick, dynamic modifications of proteins, it's useful to have small chemical units that can be quickly added on and then removed. Histone acetylation causes the nucleosome to loosen its tight grip on the DNA, so facilitating increased gene transcription.[11] Removal of the acetyl groups, known as deacetylation, causes the reverse: silencing of the genes.

It should not be thought that CpG methylation and the regulation of histone acetylation are quite separate mechanisms. In reality they are carefully coordinated. Methylated DNA regulates transcription by directing gene-silencing machinery to specific promoters, and binding proteins that recognize methylated cytosines pull in the histone deacetylases that reduce gene transcription.

The various histone modifications are reproduced in the daughter chromosomes during cell division, in parallel with the cytosine methylation pattern. The collection of histone markers is known as the "histone code." As you can imagine, reproducing thousands of such chemical markers on thousands of different histone molecules requires some exceptionally sophisticated molecular machinery, certainly more complex than the mechanism involved in reproducing the CpG methylation pattern. In fact, the molecular machinery

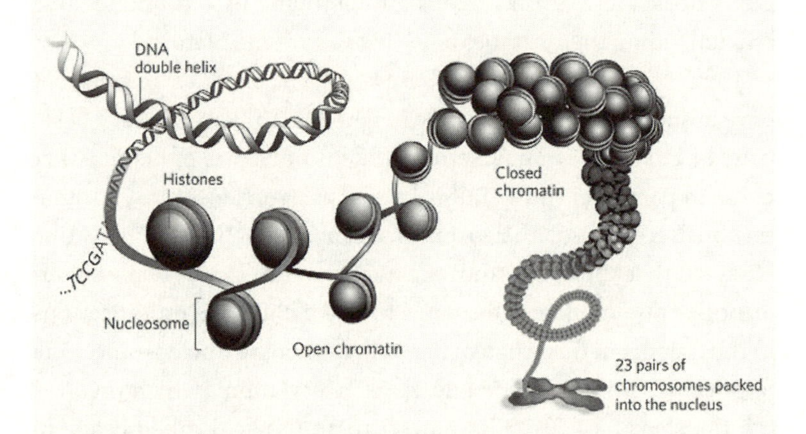

FIGURE 10.1. DNA is packaged with histone proteins to form chromatin, the central units of which are the nucleosomes, which then pack up even further to make chromosomes. Adapted by permission from Macmillan Publishers Ltd. Baylin, S., and Schuebel, K. *Nature* 448 (2008): 548–49.

is still an active area of research, and good reviews are available for those who wish to pursue the topic further.[12]

One of the most dramatic and well-studied examples of an inherited type of gene silencing is the X chromosome inactivation that occurs in mammals at an early stage of embryonic development. Remember that females are characterized by having a pair of X chromosomes, whereas males are XY. It would be biochemically unwise, and certainly unfair to males, if females made twice as much of the X-linked gene products as males. Therefore one of the two X chromosomes is shut down by massive CpG methylation, deacetylation of histones, and other mechanisms, in such a way that all the cells that make up female mammalian bodies have one of their X chromosomes switched on, but not the other.

If all this seems terribly esoteric, the results are actually quite mundane. Take a closer look at the next tortoiseshell or calico cat that you happen to encounter. The gene controlling the coat color is located on the cat's X chromosome. But the allele making the fur black is on one X chromosome, and the allele making it orange is on the other. The skin cells are a mosaic in which either one or the other X chromosome is inactivated, giving rise to the pleasant variegated black-orange coloring of the tortoiseshell. So next time you stroke one, think methylation and deacetylation; the cat will be impressed.

The third epigenetic modification besides CpG methylation and histone modifications that has gained recent attention involves microRNAs, and their role seems to be that of "directors" as they steer the methylation and acetylation molecular machinery to the appropriate genes.

Epigenetics in Action

Methylation, acetylation, and microRNA provide the molecular mechanisms. The four examples that follow illustrate the various ways in which the coordinated DNA modifications that result convey epigenetic information both within and across generations. If the biochemistry seemed a bit heavy-going in the previous section, at least it's a lot easier to visualize the consequences when you look at plants or encounter your next mouse.

A Coat of Many Colors

When people die at different ages, we are not that surprised. After all, together we represent a rich repertoire of genetic diversity. Given the range of pathogens that we all have to face during our lifetimes, plus our own richly diverse genetic heritages, it is maybe not surprising that some of us live quite a bit longer than others. But now let's take mice that have been inbred for more than thirty generations so they are essentially identical from a genetic perspective. Furthermore, these mice are fed an identical diet and maintained in an identical environment with regard to temperature, humidity, and dark-light cycles. Despite all this, there is still a substantial variation in life spans; the earliest deaths often occur at about eighteen months of age, whereas as many as 10 percent of the animals survive to thirty months or longer.[13] How can the age of death vary by as much as 40 percent in genetically identical individuals?

Random epigenetic variation in gene control seems to provide at least part of the answer. This is well illustrated by the case of the agouti gene, which has been studied in mice for over a century. The gene triggers the production of golden pigments in the hair shaft, giving mice their characteristic brown coat with "highlights," and is named after a group of South American rodents with the same kind of coat coloring. However, some alleles of the agouti gene cause mice to be completely blonde or "yellow," not the color you might associate with your average house mouse.

In the "viable yellow" allele, regulation of agouti gene expression comes under the control of a nearby retrotransposon that contains a gene promoter. Retrotransposons, as presented in chapter 4, are those copy-and-paste sections of DNA that are inserted all over the genome. Their heavy methylation ensures that they are kept quiet: no one wants random bits of inserted DNA to cause a nuisance. However, in some cases their methylation status is variable, and that in turn can affect the expression of nearby genes. That is precisely the situation with the agouti gene.

In practice, genetically identical mice with a single copy of the viable yellow version of the agouti gene can be born with coats of many different colors, ranging from fully yellow through various shades of mottling to a full agouti coat. We can imagine agouti gene expression as being controlled by a big lever. If the lever is in the fully on position (no methylation on the retrotransposon), then agouti is expressed all over the place, and you get the really yellow mice. As the lever is gradually pushed to "off" (increasing methylation of the retrotransposon), the agouti gene is steadily switched off until expression returns to the normal pattern, and the agouti coat color is restored.

As it happens the agouti gene can affect aspects of mouse physiology that are far more significant than fur color. When the agouti gene is fully switched on (no methylation on the retrotransposon), all kinds of other metabolic effects ensue; the yellow mice are fatter, tend to get adult diabetes, have increased susceptibility to cancer, and die younger. Taken together, these characteristics are called the "yellow agouti obese mouse syndrome." The agouti-colored mice, however, have none of these problems and with their viable yellow alleles well methylated, live long and no doubt productive lives.

But that is not the end of the story. Although the extent of the agouti gene expression seems to be random as far as the baby mice are concerned, which is why their coat color varies randomly, the inheritance of the phenotype is not random. The coat color of the maternal parent correlates with the coat color of her offspring, so a yellow mother produces a higher proportion of yellow offspring. Even more intriguingly, if the coat color of the grandmother mouse is yellow, then this further increases the yellowness of her grandchildren mice. But this is not the case with males, either with fathers or grandfathers. The influence on the next generation is only seen in the female line.

How can this be so? This brings us back to the question as to how genetic markings are reproduced between different cell generations. As far as CpG methylation goes, we have seen that this is relatively straightforward when it's a question of somatic cell division. But between generations, things get more complicated because gametes (sperm and eggs) are involved—the germ line. Most epigenetic modifications are erased in the germ line because the whole point is to create a clean slate in which totipotent cells are present in the very early embryo, cells that have the potential to develop into any cell of the body. It's like deleting all the software from your computer so that you can reprogram it from scratch, except that it turns out that the deletion of epigenetic markers is incomplete, which explains why some epigenetic information can be passed on from one generation to another. In the case of the agouti mice, it seems that the wiping of the slate is more efficient in the male germ line than in the female germ line, which is why the epigenetic information is partially passed on from grandmothers and mothers.

Amazingly, a simple change in diet during pregnancy can cause epigenetic changes that are highly visible in the offspring in the case of the agouti mice. The methyl groups needed for methylating cytosine can either be formed in the body or supplied in the diet in the form of compounds such as betaine and choline, which contain preformed methyl groups. Furthermore, the body can be stimulated to make more methyl groups by ingesting a diet rich in folic acid. If these methyl supplements are fed to bright yellow agouti mice during pregnancy, causing increased methylation of the retrotransposon promoter, then the agouti gene becomes less active and their offspring have mottled or normal agouti fur and are much healthier.[14] They are no longer fat, don't suffer from diabetes or increased cancer risk, and live longer.

Epigenetics and Stem Cells

It is a tribute to the conceptual power of epigenetics that we can switch from mouse fur and obesity to embryonic stem cells and still be talking about similar molecular mechanisms. Embryonic stem cells are the pluripotent cells derived from the very early embryo that have the potential to develop into almost any cell of the body. The reason they are known as "pluripotent" rather than "totipotent," the term used above, is that totipotent cells can generate a new organism, whereas pluripotent cells lack this ability since they cannot develop into tissues needed for reproduction, such as the placenta.

The possible use of embryonic stem cells for therapeutic purposes has aroused huge interest, as well as some controversy due to the use of the very early human embryo as the source of such cells. The therapeutic potential has sometimes been hyped out of proportion, and the discussion has often become characterized more by heat than by light. Accurate scientific information can promote calm, reflective conversations about the ethical issues involved.

What everyone is agreed on is that it would be far better from an ethical point of view to find other sources of stem cells that don't depend on very early human embryos. Remarkably, in 2006 a Japanese research group found that it was possible to reprogram mature mouse skin cells and convert them to cells that are very similar to embryonic stem cells, and soon after the feat was performed for human cells as well.[15] This was achieved simply by transferring into the cells a cocktail of only four transcription factors that caused the cells to wipe off their own epigenetic markers that made them skin cells and revert to their original, pristine state. Such cells are called "induced pluripotent stem cells" (iPSC). In the laboratory, further cocktails of reagents can be added to convert such cells into liver cells, skin cells, neurons, or more or less any type of cell that might be needed. If iPSCs eventually prove to be as good as embryonic stem cells for potential medical therapy, then one advantage is that they could be made from a patient's own cells, thereby overcoming any problem of tissue rejection. However, whether this would ever be practical from a time and cost point of view is quite another matter.

But the important question that remains is whether such iPSCs are really the same as embryonic stem cells. The acid test for "real" stem cells is whether they can be infused into mouse blastocysts (very early embryos) in such a way that the blastocysts can be transferred to a foster mother which then gives birth to normal generations of baby mice containing only the donor stem cells.[16] When mouse iPSCs were developed in the laboratory and then compared with mouse embryonic stem cells, it was found that their expression of mRNAs and microRNAs was indistinguishable.[17] But not quite. Just one small gene cluster on Chromosome 12 showed reduced acetylation and was not expressed as mRNA in the iPSCs, whereas it was expressed normally in the embryonic stem cells. Furthermore, these iPSCs did not manage to generate lines of mice consisting only of cells originating in the iPSCs; in other words, they were not truly pluripotent. Something was missing.

To see whether the silenced gene cluster could be the culprit, the researchers simply added histone deacetylase inhibitor to the iPSCs. This caused increased acetylation of the histones in the silenced gene cluster so that the genes were now switched back on. Remarkably, these treated iPSCs now behaved like "proper" stem cells and gave rise to generations of little mice with cells all derived from the treated iPSCs. Just adding the right chemical was sufficient to modify the genome epigenetically in such a way that it behaved like a normal stem cell genome.

Epigenetics in Plants

Although our examples so far have been from animals, it is worth noting that cross-generational inheritance of epigenetic modifications is especially common in plants. Indeed, the first epigenetic changes ever described were those that occur in maize, way back in 1956.[18] In fact, there is much more scope for epigenetic modifications in plants than, for example, in mammals.

In mammals the germ-line cells begin to differentiate from the somatic cells very early during development, during gastrulation when they develop from the so-called posterior primitive streak. This occurs from the sixth day of mouse fetal development, or from the fourteenth day for the human—providing a very narrow time window for epigenetic changes to occur in the germ line. In plants, by contrast, there is no early separation between the germ-line and somatic cells, and the gametes develop much later out of somatic cells. The environmental changes that cause epigenetic modifications in the genome thus have more of a chance of being passed on in the germ cells; the window of environmental opportunity is much wider. This difference in the pluripotency window also explains why children who plant rats' tails in the backyard are disappointed when no rats grow up out of the ground, whereas if they do the same with many types of plant cutting, they can have the satisfaction of seeing the plants reproduce themselves.

From an evolutionary perspective, the main reason that plants are keen on epigenetics is clear. Plants can't move, at least not very far. If the environment changes and an animal doesn't like it, then it can always go elsewhere. Plants cannot, and it is therefore greatly to their advantage if they can adapt to their environment in a temporary kind of way. Epigenetic modifications are useful for such short-term adaptive responses, as in responses to predators.

For example, wild radishes can be damaged by a rapacious caterpillar known, suitably enough, as *Pieris rapae*.[19] Not surprisingly, the radishes fight back, triggering the production of a really nasty-sounding group of chemicals called mustard oil glycosides and making their new leaves more hairy, in order to ward off the unwanted invaders. But what is really interesting in the present context is that the increased defense response, nasty chemicals plus hairy barrier, is also passed on to the radish progeny, so that caterpillars were 20 percent slimmer when they tried feeding on the progeny from damaged as compared to undamaged plants. Serves them right, too.

What is striking about such changes is the way that they can affect a whole plant population all at once, whereas Mendelian inheritance due to mutations in the nucleotide sequence would affect only certain individuals in the population, and would take far longer to start exerting benefits in terms of adaptive responses to a changed environment. Epigenetic change therefore provides organisms with the option for rapid adaptation to the environment in a way that can benefit their offspring and prepare them for the environmental changes that they are as likely to face as much as their forebears.

Coping with Stress: A Lesson from Rats

Rats get stressed as well as plants, and like the radish and its persecuting caterpillars, the stress response in rats is thought to represent an inducible defense mechanism.[20] What happens is quite fascinating (Figure 10.2). When more predators are around, there is less time for maternal care, which is normally bestowed by licking and grooming, together with behavior known in the ratty world as "arch-back nursing," which is what the maternal rat does when it's nursing its pups. Here we will simply call this whole panoply of activities "maternal care."

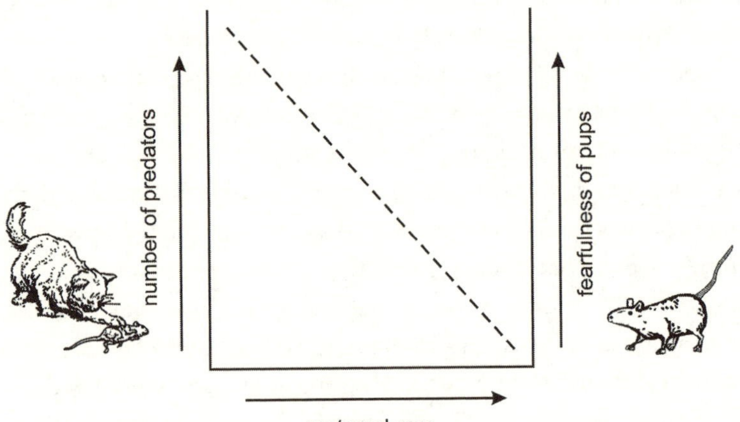

FIGURE 10.2. Rat maternal care. As the number of predators increase, rats tend to nurse in a crouched, protective position rather than a relaxed, arch-back condition. Less maternal care, particularly less time being nursed by a mother in the arch-back position, is correlated with increased fearfulness in rat pups.

With less maternal care, the offspring become more fearful and watchful. The idea is that this is a good thing for rats, because they'll be less likely to be eaten by a cat, or whatever. So the response of the pups is thought to be adaptive—it has some value for survival. By contrast, the offspring arising from a high maternal care environment are generally less fearful and more laid-back. Life for them in the rat nursery has been a rather pleasant affair. These nonfearful or fearful traits persist on into adulthood, and then on into the next generation, because the females demonstrate either fearful or nonfearful responses to their own new litters in turn. In fact, if foster pups are placed with one type of mother or the other during their first few weeks of life, then they will learn the stance adopted by their mother, ending up as either the more fearful or the more laid-back type of rat. And so life goes on.

Now simply reading this rather touching rat narrative, you might be tempted to just think that rats, like humans, are very sensitive to their experiences in early childhood, and make a mental note to be nice to babies of whatever species next time you meet them. But in fact this rat story is one about a complex interweaving of behavioral, physiological, cellular, and molecular factors.

High maternal care is associated with a broad array of changes in the rat brain that include an increased number of glucocorticoid receptors in the hippocampus. Stress responses are triggered in mammals via the steroids that bind to such receptors and mediate their effects. The hippocampus is the part of the brain especially involved in laying down the memory traces in long-term memory and memory recall, as well as spatial navigation. Signals mediated through the glucocorticoid receptors are associated with decreased CpG methylation and increased histone acetylation, leading in turn to increased gene transcription, and the production of yet more glucocorticoid receptors, a condition that remains throughout the rat's life.

Comparison of the hippocampus from the rats that previously experienced high or low maternal care has revealed no fewer than three hundred different genes that are expressed at different levels. Never underestimate the impact of mothering on the physical well-being of the offspring. Gene expression can be modulated by infusing methyl donors or histone deacteylase inhibitors into the rat brains, consistent with the idea that the differences are controlled by epigenetic modifications.

What this example highlights is the way that genetic, epigenetic, and environmental factors are woven together into a rich tapestry in the intact organism, and it would be a mistake to suggest that only one of these instruments is playing the music of life.

The Nonidentical Identical Twins

Having started with animals, worked our way through plants, and then back through rodents, it seems somehow appropriate that we should finish this section with human identical twins. Or are they? We have already noted that although identical twins start life with the same genomes, in practice they gradually accumulate somatic gene mutations during life, and can differ at the DNA nucleotide level, albeit rarely, in items such as the number of CAG repeats in their huntingtin genes.

Despite all those TV programs telling us how amazingly similar twins are in so many ways, and indeed they often are, what is perhaps more surprising is how very different they can be as well. Detailed comparison of the "epigenome" of identical twins has revealed some interesting differences. The epigenome is the profile of gene methylation and histone acetylation across the genome that provides a good indication of how many genes are expressed or silenced. In one study involving eighty identical twin pairs from Spain, about a third of the cohort were found to display epigenetic differences between the twin pairs, distributed throughout their genomes.[21] These differences were especially marked in twins who had lived apart the longest and who had adopted different lifestyles. Smoking, exercise, and diet are among the many changes that can trigger epigenetic modifications. The transfer of epigenetic information through successive cell divisions is also imperfect, as already noted, and so a random process of "epigenetic drift" can also lead to a divergence of epigenomes with the passage of time. Such changes, taken together, likely contribute to the differences in aging, health, and longevity that are seen between identical twins. Cloned animals, in which the DNA has a single parental source, likewise display considerable epigenetic differences.

In fact, perhaps surprisingly, having identical genomes does not seem to make much difference to the degree of variation between identical twins when it comes to longevity (age at death) in comparison with pairs of sibs. In one study of 184 pairs of twins, a comparison was made between the identical (monozygotic) and nonidentical (dizygotic) pairs with respect to their age at death.[22] Whereas the identical twins come from the splitting of a single zygote, so they share the same genome (and therefore of course must be of the same sex), the dizygotic twins come from the fact that two separate eggs were fertilized simultaneously, so the two zygotes that result are genetically no more similar than any two sibs, and they may be of a different sex.[23] In the Spanish study, the difference in the age of death between the monozygotic twin pairs was seven years on average, but such averages hide the fact that the age differences ranged from a couple of weeks to eighteen years. In the case of the nonidentical twins, the difference in age at time of death was nine years, and the range was three to nineteen years. So there was really not that much in it.

Other studies on large cohorts of twins support the conclusion that the genetic contribution to the variation in age of death is significant but rather modest.[24] One group of Swedish researchers compared longevity in 3,656 identical and 6,849 nonidentical (sex-matched) twin pairs and concluded, "Over the total age range examined, a maximum of around one third of the variance in longevity is attributable to genetic factors, and almost all of the remaining variance is due to non-shared, individual specific environmental factors."[25] Of course, those environmental factors are very likely to involve a considerable degree of epigenetic modification of the genome. Earlier studies on the nature-nurture question tended to assume that it was either one or the other. We now know that nature and nurture are so thoroughly interconnected that assigning causality to either one or the other has become less meaningful.

Epigenetics and Disease

From our discussion so far, it will be no surprise at all to hear that epigenetics and disease are intimately related. Could it be that environmental factors induce epigenetic changes that in turn lead to disease?

Alzheimer's Disease

Alzheimer's disease, named after the German physician Alois Alzheimer in 1906, is a progressive dementia that generally doesn't begin until after the age of sixty-five, though there is an early-onset form that develops in the thirty-to-sixty age range. The early-onset form accounts for less than 1 percent of the 27 million cases of Alzheimer's documented worldwide (in 2006).[26] Due to the aging populations of the world, Alzheimer's is predicted to affect one in eighty-five people by the year 2050. Apart from the 5 percent of cases that show a classic dominant type of inheritance due to well-defined gene mutations, the remaining 95 percent of cases are "sporadic," though there is clearly some familial susceptibility involving the contribution of multiple variant genes.

In one identical twin study, two male chemical engineers had died, one from Alzheimer's disease at the age of seventy-six after the symptoms of the disease had started to emerge at age sixty.[27] This engineer had been exposed extensively to pesticides during the course of his work. His engineer twin had done a different kind of work, not involving pesticides, and died at age seventy-nine from prostate cancer. With the usual ethical permissions in place, brain samples were obtained from both twins very soon after death. What they showed was that neurons (brain cells) taken from a specific part of the brain known as the neocortex, a part severely affected in Alzheimer's patients, displayed much lower gene CpG methylation levels in the sample from the Alzheimer's twin when compared to his genetically identical brother. This in turn could affect the expression levels of thousands of different genes, and indeed mRNA levels vary considerably for different genes in the Alzheimer's brain. The big challenge is to sort out the really key differences that trigger the disease in the first place, but it seems a reasonable assumption that epigenetic as well as genetic variation are both involved.

Highly relevant to the findings in Alzheimer's patients, in whom memory loss is a disturbing and significant clinical sign, is the observation that specific gene methylation changes occur in the brains of mice that are laying down new memory traces.[28] Disruption of such mechanisms resulted in memory deficits in mice so they were unable to remember recently acquired tasks. Epigenetic changes appear to be intimately involved in our ability to remember things as well as our forgetfulness.

Moreover, the insight that most genes identified to date that cause mental retardation are involved in the processes that lead to epigenetic change in itself points to an important role for epigenetics in brain function and dysfunction.[29]

The Epigenetics of Cancer

Some of the most striking examples of epigenetic involvement have been found in cancers, not least the ability of the environment to induce epigenetic changes that increase cancer susceptibility. The World Health Organization has estimated that more than 13 million deaths annually are due to environmental causes, of which 24 percent relate to preventable exposures.[30] A recent report revealed that 148 different chemicals, many of them carcinogens (cancer-inducing agents), were found in samples of blood from a U.S. population.[31]

An animal study helps to focus the mind on the impact of environmental chemical reagents with respect to the health of future generations. Female pregnant rats were exposed to vinclozolin, a common fungicide used in vineyards, and the health of the offspring was followed through four generations.[32] Increased disease states in 90 percent of the offspring were noted right across the four generations of males, but not females, including increased tumor incidence, prostate disease, kidney disease, and abnormalities of the immune system. Note that only the first rat generation of pregnant mothers was exposed to the fungicide. The transgenerational effects correlated with epigenetic changes in the male but not female germ line. The sex differences in the induction and transmission of epigenetic mutations should not be surprising because, as already noted, reprogramming of the epigenome occurs at different times in the testes and ovaries, and sex differences are also involved in the efficiency of erasure of the old marks. Not all slates are equal.

Translating the results of such animal studies into the human context is not easy, but several human studies at least strongly suggest that disease states can be promoted across generations by means of epigenetic mechanisms. For example, a form of colorectal cancer runs in families that is associated with the abnormal methylation and therefore silencing of two DNA repair genes that are normally kept fully switched on to protect the genome.[33] The same phenomenon has been detected over several generations of the same family, suggesting that this may be an inherited epigenetic modification.

A generalized reduction in gene methylation can exert a destabilizing effect on the genome, causing increased cancer risk. This effect is seen in cells or animals lacking the CpG-methylating enzymes. Hypomethylation of DNA (meaning a reduction in DNA methylation) is also observed in many tumors, particularly in areas of the genome that don't have many protein-coding genes. One reason for genomic instability due to decreased methylation is the 45 percent of the genome consisting of transposons, copy-and-paste DNA sequences. As already mentioned, these are generally kept quiet by blanket methylation. But if their demethylation occurs, then copy-and-paste activity increases and more transposons arrive in the genome, causing increased instability of the genome in turn, involving various types of chromosomal rearrangement of the kind described in chapter 4.

One of the penalties of aging is a gradual reduction in transposon methylation levels, which may contribute to the fact that cancers are generally associated with older rather than younger individuals. It is not just aging that accelerates this process, but environmental factors as well. In one study, the exposure of 1,097 men living in the Boston area to black carbon derived from vehicle emissions was found to be correlated with DNA methylation levels in one of their classes of transposon—the more pollution, the less methylation, meaning more genomic instability.

Fortunately every cloud has a silver lining, and better insights into the epigenetics of cancer are also opening up new avenues for therapy. Once the molecular mechanisms involved are understood, then rational drug design becomes feasible.

Genetic Imprinting and Disease

The phenomenon of genetic imprinting also provides the key to understanding a cluster of rare epigenetic diseases. Normally genes can be transcribed in an equivalent manner from either chromosome of a pair. If it were not so, then diseases would not be inherited in a heterozygous manner, because one mutant gene would not be compensated for by the product of the normal gene on the other chromosome. In other words, all diseases caused by a single mutant gene would be inherited in a dominant manner. Fortunately this is not the case: we should all be thankful for our backup chromosome. (Always maintain a backup of your hard drive.)

However, not all genes can be transcribed in this way from either chromosome, at least in mammals and flowering plants. Mammals have about a hundred or more genes that are imprinted at birth, which means that either the paternally or maternally derived genes are switched off by methylation and other mechanisms, and this silencing is then maintained throughout the subsequent production of cells in the lifetime of the organism.[34] Imprinting is nonrandom: the same sets of genes are silenced consistently on either paternally or maternally derived chromosomes.

Not surprisingly, aberrant imprinting can lead to disease. A mutation in a gene on one chromosome that might have been compensated for by the normal gene on the other chromosome is now free to express its damaging effects. This is the case with Angelman syndrome and for Prader-Willi syndrome. Both can be produced by a similar deletion of a small portion of Chromosome 15, but which syndrome develops depends on whether the deletion is present on the maternal or paternal chromosome, for the segment contains different sets of genes that are silenced on either one or the other chromosome.

The example of Angelman syndrome illustrates the point. The condition results from a failure of normal development of the nervous system, leading to jerky movements, frequent seizures, sleep disturbance, and yet despite all these challenges, often a happy and smiling outward demeanor. Those afflicted with the syndrome do not generally develop more than five to ten words, if any. Dr. Harry Angelman was the physician from northern England who first encountered three children in his medical practice with the condition, although at first he was unsure whether it was a single condition or several. Only after he went on holiday to Italy and saw an oil painting in the Castelvecchio Museum in Verona of a smiling boy entitled a "Boy with a Puppet" (Figure 10.3) did Dr. Angelman realize that perhaps this condition went back centuries, in turn leading him to write up his description for a medical journal.

Normally the genes on Chromosome 15 involved in the deletion that leads to Angelman syndrome are expressed only in the

Figure 10.3. "Boy with a Puppet" by Giovani Francesco Caroto (1480–c. 1555). This painting inspired Dr. Harry Angelman, who first diagnosed the disorder now known as Angelman syndrome. The boy's rigid smile reminded Dr. Angelman of his patients' jerky movements. It was later discovered that around half of children with Angelman syndrome are missing part of Chromosome 15.

maternal copy of the chromosome, as the relevant paternal copies of the genes from this region are in any case silenced by imprinting. When the deletion occurs in the maternal chromosome, there is no backup, and the neurodegenerative consequences begin to appear. If any offspring ensue, then the syndrome is therefore inherited as a dominant condition.

Various interesting theories abound about why imprinting might have evolved in the first place, given the negative outcome from mutations that wouldn't have made any difference had the usual backup copies been available. The competing theories, still rather speculative anyway, lie beyond the scope of this chapter,[35] but all suggest ways in which the evolutionary fitness of particular individuals might have been increased by imprinting mechanisms. Evolutionary history in general turns out to be a story about the acquisition of selective advantages that so often carry various costs along with them.

Epigenetics, Lamarck, and Evolution

We referred in chapter 1 to the fact that Darwin was a mild Lamarckian, believing that to some extent the inheritance of acquired characteristics played some role in the evolutionary process. However the variation comes into the phenotype, natural selection still operates by providing selective advantages to those organisms that are fitter for their environments. But the clear separation between somatic and germ-line cells established by August Weismann and others in the closing years of the nineteenth century did make Lamarckian mechanisms seem much less likely, and even less so following the emergence of the neo-Darwinian synthesis in the 1930s and onward.

Yet repeatedly in biology, just as Lamarck seems to have finally been shown out the front door, he seems to creep in again through the back. For some reason the possibility that acquired characteristics in any shape or form might be inherited has often aroused great passions among some biologists, as if the merest whiff of Lamarckian thought represents some terrible heresy that will subvert the true path of Darwinian rectitude. Since Darwin himself was clearly Lamarckian in some of his thinking, this seems a bit odd. Scientific theories are like maps that incorporate different kinds of data and render them coherent. Just as maps are revised to incorporate new data as it becomes available, so no scientific theory is static and should be revised as and when necessary.

Certainly the history of attempts to reinstate the inheritance of acquired characteristics into mainstream biology has seen some murky episodes, which itself has made the topic a controversial one. The discussion has not been helped by the ways in which the science has so often become mixed up with ideology.

In 1932 Trofim Lysenko became a leading agriculturalist during the early years of the Soviet Union, his rise favored by Stalin who was looking for a "quick fix" to save his country's agriculture following the disastrous famines that arose from enforced collectivization. Rejecting Mendelian laws of inheritance and the idea of the gene, Lysenko and his followers drew on the popular ideas of Marxism-Darwinism current at the time to claim that external conditions could directly affect the heredity of plants, proposing this as the scientific basis for the whole of Soviet agriculture.[36] Genetics was portrayed as Western, bourgeois, capitalist, and Christian, the latter objection playing on the fact that Mendel was a monk. The geneticists in turn accused Lysenko of "Lamarckism." By 1948, Lysenko had managed to gain control over all the country's biological institutions, leading to the eclipse of genetics and evolutionary studies, a stranglehold that was not finally broken until the mid-1960s. What is so striking about this sorry episode is the way in which all the competing parties claimed that they were the true standard-bearers of Marxism-Darwinism, adjusting their scientific claims to fit with current Soviet government rhetoric. Small wonder that "Lamarckism" eventually ended up becoming a term of abuse.

A very different example that also had the effect of portraying

Lamarckism as villain comes from the curious tale of the Midwife Toad.[37] The story centers on an Austrian biologist named Paul Kammerer who in the early 1900s claimed that he had demonstrated Lamarckian inheritance in various animals, including salamanders. At the time, his reports were dismissed and he ended up committing suicide. More recently, however, Alexander Vargas, a developmental biologist at the University of Chile in Santiago, has attempted to restore his scientific reputation, pointing out that epigenetic effects could explain Kammerer's results, maintaining that the case represents "a tragic case of scientific incomprehension."[38]

Given these and other examples in the history of science, it's perhaps not surprising that the possibility of Lamarckian inheritance still arouses such passions among biologists. But at the end of the day, much of the discussion comes down to semantics. Lamarck himself envisaged that the central mechanism in evolution involves the gradual adaptation of animals and plants to the environment by means of the use or lack of use of various organs. The more frequently and continually an organ was used, the stronger it would become, and this better organ would then be passed on to the progeny. Conversely, if an organ was little used, then it would gradually be removed as the generations proceeded. We know this central Lamarckian idea to be wrong, and Darwin replaced it with the idea of natural selection, but nevertheless retained as a supplementary mechanism the Lamarckian idea that acquired adaptations could be passed on to progeny.

Epigenetic modifications regulating gene expression that are induced in the first generation by a particular environmental change and then passed on across several generations are not really the kind of long-term evolutionary changes that Lamarck had in mind. Epigenetic changes do not generally seem to be transmitted for more than a few generations, whereas the genome sequence carries on over hundreds and thousands of generations. On the other hand, there seems to be no reason in principle why epigenetic changes in response to the environment should not play some role in evolutionary change. If radishes become more resistant to prowling caterpillars for several generations, their fitness will increase, and the particular sets of genomic variants that they currently possess are therefore more likely to be passed on to succeeding generations. If those sets of genomic variations are particularly susceptible to the kind of beneficial epigenetic modifications that occur in response to greedy caterpillars, then such sets will give future radishes a better chance in life. In other words, their fitness will increase. So the evolution that occurs looks quite Darwinian after all: it is natural selection that has the final say.

The jury is still out as to whether epigenetics really makes that much difference to evolution. But since epigenetic changes provide the potential for organisms to respond to environmental changes in a flexible and, after a generation or so, reversible manner, it would be surprising if such benefits did not contribute in turn to longer-term reproductive success encoded in the genomic sequence.

The Era of Epigenomic Sequencing

The epigenomic information available so far from different species is merely the tip of a very large iceberg. With more than two hundred tissues, each one with its own unique epigenomic signature that changes through time, the "space" of human epigenomics is vast indeed when compared with the "space" of genomic information. Furthermore, the question also arises as to when the "epigenome" is complete. About 56 million CpG sites reside on the forty-six human chromosomes, and many methylated cytosines lie outside the "CpG islands." So a proper "methylome" will need to examine all of the billion-plus cytosines in the human genome. But as we have already seen, cytosine methylation is only part of the story. What about chromatin modifications? The four core histone pairs that, together with DNA, constitute chromatin, contain altogether more than one hundred amino acids at which modifications (such as acetylation) can change the way that histones regulate gene expression.

Despite all these challenges, the ambitious project of sequencing the human epigenome was formally launched in 2010 under the auspices of the International Human Genome Consortium.[39] As with DNA sequencing, new technologies have rendered the project feasible, although it is still sometimes referred to as the "Large Hadron Collider" project of biology. The cost is not as astronomical, but it's nevertheless a major project, and the data generated are likely to have a significant impact on biology and medicine for many years to come.

CHAPTER 11
Genetic Engineering

IN SOME WAYS the term "genetic engineering" is an unfortunate one, as it conjures up methods dependent on artificial materials and the kind of engineering techniques that we're familiar with in building cars, airplanes, and so forth. The wicked scientists who create monsters in horror movies generally seem to be surrounded by lots of wonderful machines, giving the impression that genetic engineering is a very high-tech kind of enterprise.

In reality the opposite is the case. The traditional techniques and materials used by molecular biologists to manipulate DNA are mostly taken directly from the natural world, where they have precisely the same functions, albeit used in different contexts. And humankind has been involved in genetic modification for thousands of years, even though the mechanisms were not understood at the time. In chapter 1 we mentioned the Assyrians and Babylonians who manipulated genes when they pollinated their date palms, not knowing a thing about genetics. In chapter 6 likewise we cited the example of the monks breeding carp with fewer scales in the garden ponds of their medieval monasteries, in the process breeding fish carrying mutations for the duplicated copy of the gene *fgfr1*.

The breeding of different strains of wheat provides another classic example of genetic manipulation. Our modern wheat strains are certainly very different from the wild grass from which they originally derived. Domestication of wheat has been traced back to around 9000 BCE in early Neolithic settlements of southeast Turkey. The *emmer* and *durum* wheat strains of those early years have undergone intensive selection and hybridization with other grass species in order to generate the robust, high-yield wheat strains that help feed a hungry world today. During the twentieth century, with the increasing understanding of genetics, crop-breeding decisions became better informed, but even without that understanding of the mechanisms involved, thousands of years of breeding had already been highly successful in terms of increased yields.

A bit closer to home for some of us is the incredible array of domestic dog breeds, all of the same species, but morphologically highly distinct. The sequencing of the dog genome and detailed comparative analysis have revealed that only a small handful of genetic variants generates this impressive diversity. Dog breeders have been altering the dog germ line now for centuries, albeit not by very much as it turns out.

So genetic manipulation is nothing new, and genetic engineering should be seen as a set of techniques that dramatically speeds up what to some extent humanity has always been doing in the biological world, albeit on a relatively slow scale. Having said that, it is also fair to point out that human genetic manipulation is now bringing about changes well beyond those that normally occur during the evolutionary process.

The central player in the story is recombinant DNA—DNA that is "recombined" from more than one source. A standard set of biological tools is required to make recombinant DNA, and it's worth understanding something about those tools because they crop up all over the literature on genetics.

The Genetic Engineering Tool Kit

Restriction Enzymes

The first item needed in the tool bag is a good set of scissors in order to cut the DNA at precise points. This basic bit of equipment is provided by enzymes purified from bacteria that are known as "restriction enzymes." They originally derived their name from the fact that certain bacteriophages (viruses that infect bacteria) were found to be "restricted" to infecting some strains of bacteria but not others. Only later was it found that this "restriction" was due to enzymes that cut DNA at specific sites in the nucleotide sequence. The reason that bacteria contain these enzymes is in order to defend themselves against attack from viruses. Since viruses contain either DNA or RNA, the best form of defense is to chop them up quick before they cause any trouble.

More than three thousand restriction enzymes that can cut DNA at more than one hundred specific sequences have now been described, and six hundred or so of these are available commercially. The ones used for genetic engineering cut through both strands of the DNA double-helix at specific sequence sites four to eight base-pairs long that are usually "palindromic," meaning that the sense and antisense strands of the DNA read the same, but in different directions. For example, a restriction enzyme called *Hin*dIII[1] cuts the double-helix as shown in Figure 11.1.

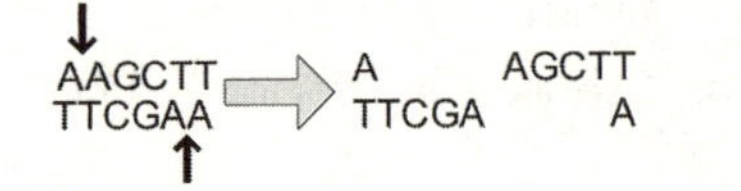

Figure 11.1

The two arrows on the left show where the "scissors" cut, and the resulting pieces are shown on the right. Conveniently this restriction enzyme, and many others, creates overhanging "sticky" ends. In other words, these ends are complementary to each other and are cohesive. If you create a mix of lots of different random DNA fragments, then two bearing these complementary sticky ends are bound to find each other in the general swamp of molecules. Another enzyme called a DNA ligase can then be used to seal the two ends together. Once again, this enzyme was originally isolated from bacteria. Cells have been doing genetic engineering for billions of years; we are just catching up!

The different sequences recognized by different restriction enzymes occur randomly throughout the genome, and each restriction enzyme creates its own unique pattern of any genome that it digests, as Figure 11.2 illustrates. The longer the sequence that the enzyme requires for recognition, the fewer DNA pieces that are created, because the chance of that precise sequence appearing in the genome will be lower. The DNA fragments can be separated rather easily on a material called agarose through which an electric current is passed. As the pieces are pulled through the gel by the current, they separate out according to their size, and a "restriction map" can be generated from the pattern observed. Each restriction enzyme creates its own unique digestion pattern because it recognizes different sequences for cutting. If a variant base-pair is located within a recognition site, then the enzyme will no longer recognize that site, and a longer than expected piece of DNA will be found on the agarose gel. Such analysis can be useful when analyzing DNA for the presence of variations.

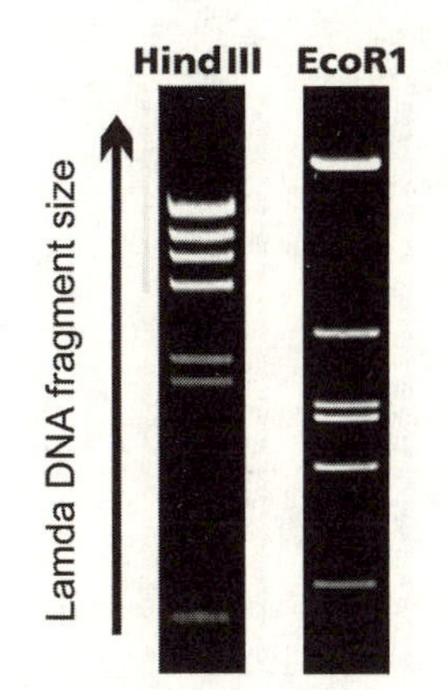

FIGURE 11.2. Restriction enzyme cutting patterns. These photographs show the different sizes of DNA segments that are produced following cleavage of the same DNA by two different restriction enzymes, HindIII and EcoR1. The cleaved DNA segments are separated by size in a process called electrophoresis. Each enzyme cuts the DNA at different sites, producing a pattern characteristic for that enzyme. Reprinted by permission from GeneOn.

Vectors

The next step toward making recombinant DNA is to find a transport system that will carry a DNA fragment containing one or more genes into another cell. The transport system is known as a "vector." Two types of vector are commonly used—plasmids and phages—as illustrated in Figure 11.3, and once again both have been around for billions of years.

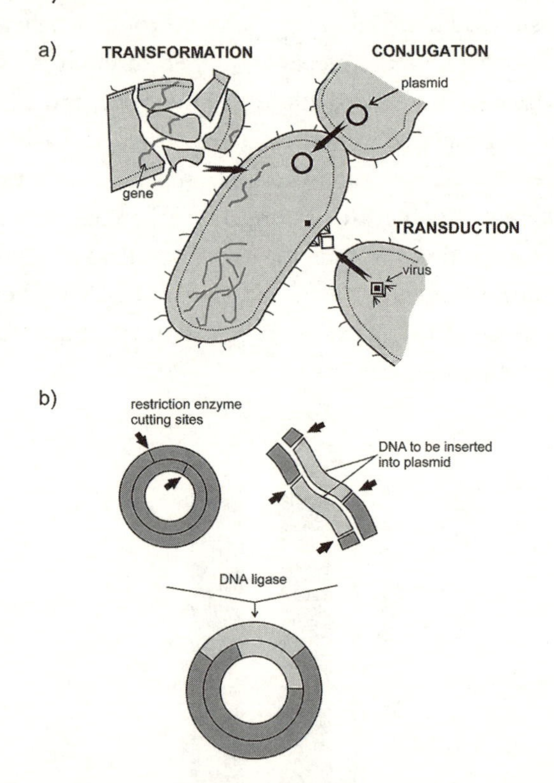

FIGURE 11.3. Recombinant DNA. Genetic engineering depends heavily on techniques that allow for the selective expression of certain genes, most commonly in bacteria. a) DNA can be introduced into bacteria by transfer of plasmids into the bacterium (conjugation), infection by a virus known as a bacteriophage (transduction), or by direct transfer of DNA into the cell (transformation). b) Recombinant plasmid DNA can be generated by cutting a plasmid with a restriction enzyme and mixing the cut plasmid with complementary DNA fragments, followed by rejoining of the segments using an enzyme called DNA ligase.

We encountered plasmids in chapter 4 in the context of lateral gene transfer in bacteria, and in particular the transfer of multidrug resistance from one bacterium to another. Plasmids are circular minichromosomes that exist in bacteria, often in multiple copies, in addition to the bacterium's single main chromosome. Due to their small size, they can readily be separated from the main chromosome and purified for genetic manipulation in the laboratory. Their small size has another advantage: plasmids often contain only a single cutting site for a particular restriction enzyme, generating a single strand of double-helical DNA with sticky complementary ends at either end. This makes it very simple to "splice" a foreign piece of DNA with the same complementary ends into the plasmid to create recombinant DNA, joining the ends together using the ligase enzyme mentioned above.

If the aim of making the recombinant DNA is to generate the foreign gene's protein product, then a promoter needs to be incorporated that the bacteria will recognize, together with a terminator that marks the end of the gene. Once these are in place, the plasmid with its extra gene incorporated is now "transfected" into the bacterial strain of choice. This procedure can be carried out using chemicals that make the walls of the bacteria temporarily leaky, so allowing the plasmid DNA to enter the cells. Another method is "electroporation" whereby the cells are briefly shocked using an electric field that likewise generates small (transient) holes in the bacterial walls. Since the transfer efficiency is not always that great, an antibiotic resistance gene can also be incorporated into the plasmid, and then the bacteria are grown in the relevant antibiotic. Only those cells that have taken up the plasmid with the resistance gene survive. Once the plasmid is inside the cells, the bacteria's own replication machinery generates as much of the plasmid DNA as required, including the "foreign" gene that it contains.

Let's say the gene involved is the human gene that encodes insulin, the hormone required for treating people suffering from type 1 diabetes. The host bacteria might well be *Escherichia coli* (*E. coli*), which contains hundreds of plasmids in each cell so that every time a cell divides, the amount of plasmid multiplies very quickly. *E. coli* is a normal inhabitant of the human gut and can cause disease, but the strains used for genetic engineering have been modified to render them harmless. Thanks to the fact that the genetic code is universal and that essentially the same mRNA and protein production machinery are used throughout living things, the bacteria can now start to make human insulin. Native *E. coli* contains enzymes that would quickly degrade the insulin formed, but the strains used for protein production have had these attackers artificially removed. The bacteria of course do not "know" that they are harboring a human gene. All the genes in their plasmids, or in their main chromosomes for that matter, will be treated the same. For pharmaceutical purposes the bacteria are cultured in large vats on an industrial scale to make millions of bacteria, and the final stage then involves the purification of the insulin. Yeast are also used as host cells for insulin production since the final purification stage is more efficient than in bacteria.

If the purpose of growing the plasmid-containing bacteria is to produce the recombinant DNA itself, rather than its protein product, then the promoter and terminator can be left off the two ends of the gene of interest. Once the bacteria have replicated enough times in culture, the plasmid DNA is then purified and the same restriction enzyme that was originally used to insert the foreign gene into the plasmid can now be used to cut it out. The gene of interest can in this way be amplified up millions- or billions-fold. This is known as "cloning" a gene, making multiple copies of it, which should not be confused with cloning cells, or indeed cloning people (= reproductive cloning). The problem about the word "cloning" is that it has many different nuances of meaning in biology. Basically the word in biology (derived from the Greek for a "branch" or "trunk") means "making a copy of," but of course what you're making a copy of makes all the difference, leading to endless confusion when the media reports on the latest research discoveries.

Plasmids are good for incorporating lengths of DNA up to 15 kilobases (kb) long, but tend to become unstable when larger sections are incorporated. For bigger pieces of DNA up to 23 kb in length, bacteriophage provide an alternative vector. Bacteriophage, or "phage" for short, are viruses that attack bacteria and are one of the most common types of biological entity on earth. Early accounts of the healing properties of the Ganges River in India may have been due to the presence of bacteriophage in the water that attacked pathogenic bacteria, such as those that cause cholera. From the perspective of genetic engineering, a useful aspect of phage is that they already have the molecular machinery necessary to make their way through the bacterial cell wall and deliver their packages of DNA into bacterial cells. So the gene of interest is first incorporated into the phage using exactly the same kind of techniques as for plasmids, then bacteria are infected with the phage, which in turn incorporates its DNA into the host DNA, and the bacteria are grown as before.

Of course, gene jockeys always want to handle bigger and bigger pieces of DNA, so they have developed a range of further vectors to make this possible. "Cosmids" result from taking a bacteriophage, removing its own DNA, and then incorporating a plasmid in its place. The cosmid therefore uses phage's normal mechanisms to gain entry into bacteria, and can handle insertions of foreign DNA up to 44 kb in length. For those who want to handle really huge pieces of DNA, yeast artificial chromosomes (YACs) are the answer. As the name suggests, these vectors come from modifying yeast chromosomes in such a way as to increase their stability and allow the incorporation of foreign sections of DNA more than 1,000 kb long (that is, more than a million base-pairs in length).

In research, biomedical, and agricultural contexts, plant or animal cells are the ones that need to be genetically manipulated, and a wide range of other vectors and techniques are used to handle different types of cells. But the basic steps are the same: cutting, sticking, conveying, and growing recombinant DNA, all completely dependent upon the consequences of billions of years of evolution. Any good genetics textbook provides further details for those interested in reading up on the techniques, and dozens of commercial companies sell all the products needed to make and manipulate recombinant DNA. Plasmids and other vectors are sold complete with their restriction maps, and every tool that a molecular biologist might need is there waiting in the catalog to be pulled off the shelf—at a price.

In addition, one method is so ubiquitous that it's worth a mention here: the polymerase chain reaction (PCR), first developed in 1983. Once again, it depends on natural products derived from the biological world, but the nifty little machine that carries out the PCR reactions is anything but natural. About the size of a laptop in area but many-fold deeper, PCR machines have become an essential piece of kit in any molecular biology, diagnostic, or forensics lab. I'd guess that in the square mile or so within which I'm sitting here in Cambridge (U.K.), one would easily find a thousand PCR machines, probably more.

The basic job of a PCR machine is to make lots of copies of a gene from a very few, or even one copy. The machine accomplishes in a few hours what used to take days using the method of cloning genes in bacteria as described above. When DNA replicates in a cell nucleus, the process is catalyzed by an enzyme called DNA polymerase. The PCR machine uses the abilities of this enzyme to replicate genes in little tubes. This process requires a single-stranded DNA to act as a template for the polymerase plus a primer section of DNA to which nucleotide bases can then be added in a systematic manner, just as they are inside the cell.

The PCR reaction then requires three basic steps, as Figure 11.4 illustrates. First a starting solution of double-helical DNA is heated to 90 to 100 degrees centigrade in order to break the double-stranded double-helix apart to generate single-stranded DNA. In the second step, with the primer present, the temperature is rapidly reduced to the range of 30 to 65 degrees centigrade for a minute or so. This is enough time for the (predesigned) primer to stick to one end of the DNA, but not enough time for the single-stranded DNA to "anneal" back together with its partner strand. In step three the temperature is adjusted to 60 to 70 degrees centrigrade, the temperature at which the DNA polymerase can then get to work to attach the nucleotides, also pre-added to the solution, to the growing complementary chain of DNA. By the end of the cycle, a complete replication of one double-helical molecule of DNA has taken place. Each cycle only takes a few minutes.

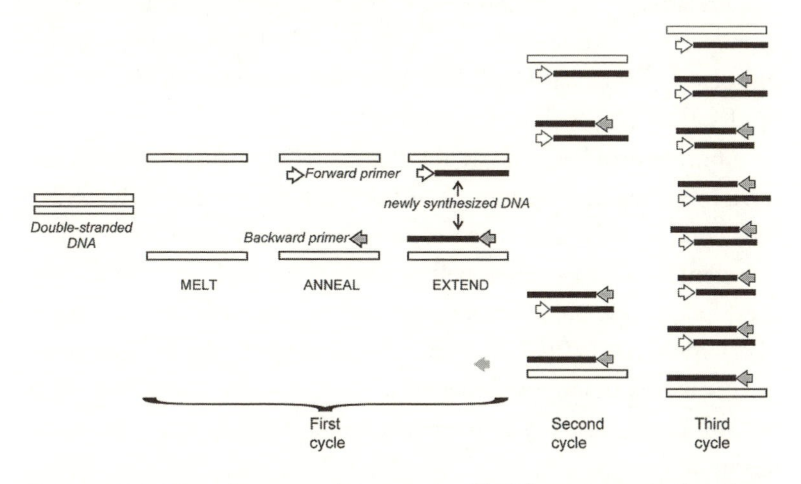

FIGURE 11.4. The polymerase chain reaction (PCR) revolutionized molecular biology. The process takes advantage of a heat-tolerant enzyme that synthesizes complementary strands of DNA to produce very large quantities of a specific DNA sequence in a matter of hours. Short DNA sequences called primers bind to complementary sequences, and the DNA sequence between two primers on opposite strands of DNA is then amplified by multiple rounds of heating and cooling.

The particular polymerase generally used is known as Taq polymerase, isolated from *Thermus aquaticus,* bacteria that live in the boiling springs of Yellowstone National Park. Since this enzyme has evolved to remain functional even in boiling water, it is the ideal choice to withstand the first step of the process, in which the DNA is separated into two single strands by high temperatures. One can simply program the whole sequence into the PCR machine to be repeated for as many cycles as required, although eventually the nucleotide ingredients for the reaction will run out unless replenished. If we start with one molecule of DNA, it will amplify up to more than a billion molecules after only thirty cycles.

As with any scientific technique, PCR has some drawbacks. One problem is that of contamination. If any unwanted DNA that matches the primer is present, then that also will be amplified up along with the rest, so a very minor DNA contaminant can make up a relatively large proportion of the final product. This becomes highly relevant to certain types of forensic DNA analysis. If the innocent person's DNA is amplified up equally with that of the guilty, then clearly there is a problem.

Another problem with PCR is that, unlike the situation in a living cell, DNA repair enzymes are not present. Native Taq polymerase does not have the ability to proofread the newly synthesized complementary DNA strands as they are synthesized. For this reason, new heat-stable polymerases with proofreading abilities have been isolated for use in PCR machines. Without them, the PCR reaction incorporates an incorrect nucleotide in about one out of every 250 to 500 base-pairs, which for many applications is not sufficiently accurate.

The PCR machine has an enormous range of uses. Besides its application in research and forensic laboratories, it is also used, for example, as a diagnostic tool to test for the presence of viruses in blood samples. The HIV virus that causes AIDS can readily be detected in this way.

Some of the tools in the genetic engineer's tool kit, then, can be quite sophisticated and high-tech, dependent upon programmable software for their operation. But even these, such as the PCR machine, are just rows of little tubes containing samples of reagents that go way back into deep evolutionary time. Without a long evolutionary history of testing out chemicals in the workshop of life, there would be no genetic engineering.

Making Synthetic Life?

If scientists can now do so much in the lab in terms of manipulating DNA and moving genes from one type of cell to another, then does this mean that they could create life synthetically in the laboratory? In principle there seems no reason why not. After all, living cells are made from DNA, RNA, proteins, and lipids organized in a certain way, so if one copies that "certain way" brick by brick, as it were, then the resulting entity would be a living cell.

The problem comes, though, with the term "synthetic life" and what that really means. The problematic status of this term was highlighted in 2010 when Craig Venter and his colleagues announced the creation of the first synthetic cell in the laboratory.[2] The news was greeted with great excitement by the media, which, as usual on such occasions, hyped up the story more than it perhaps deserved, even though the paper indeed represented a great technical achievement. After the results had been published, I remember spending a couple of hours on a Sunday morning sitting in the studio of our local radio station as I was hooked up for interviews on the topic with eight different BBC local radio stations all around Britain, one after the other. The questions indicated some confusion about what precisely had been reported.

The analogy that I found most readily communicated the achievement of the Venter group comes from the world of computers. Imagine you had a laptop in your possession, and being a clever computer engineer, you were able to remove its hard drive, take it to bits, and find out how it worked. Based on that knowledge you then constructed a new hard drive from scratch (actually this is pretty much what companies in various countries do anyway when they want to find out about their rivals' models). You then take your newly constructed hard drive and place it into a closely related laptop, albeit a slightly different model (nothing as different as a Mac and a PC). You then switch it on, and lo and behold, the newly constructed "synthetic" hard drive works in its new computer environment!

The relevance of this analogy becomes apparent as we take a closer look at what the Venter group actually accomplished. The team sequenced the million-base-pair sequence of the genome from a bacterium called *Mycoplasma mycoides*, bacteria that cause a contagious lung disease in animals such as cows and goats. *Mycoplasma* are the very smallest genus of bacteria around, lacking a cell wall, and are only about 10 percent the size of *E. coli*.

Once the *M. mycoides* genome was sequenced, the research team was then able to synthesize the complete genome from scratch. In fact, like any sensible, well-funded molecular biology lab, they went DNA shopping. Why do the boring part when you can pay companies to do the job? So they bought more than 1,000 DNA segments, each 1,080 base-pairs long, to cover the complete length of the *M. mycoides* genome. Four of the ordered segments contained, in code, an e-mail address, the names of some of the researchers involved in the project, and some famous quotations. These "watermark" sequences were included so that the synthetic DNA could be distinguished from other DNA at a later stage. The researchers then stitched together the DNA segments to generate an artificial chromosome containing the complete *M. mycoides* genome, using yeast as a vector before transferring the artificial chromosome into a closely related species of bacteria called *M. capricolum* from which its own single chromosome had been removed. After correcting for a single-point mutation, which at first subverted the whole experiment, the system eventually worked: the reassembled chromosome enabled the *M. capricolum* to grow and divide just as well as it did with its original genome. But the bugs were now making proteins characteristic of *M. mycoides* rather than *M. capricolum*: the new "hard drive" was using all the "computer's machinery," but its slightly different information was now generating a slightly different repertoire of outputs from those that were there before.

Does this represent "synthetic life"? Hardly. The genome was copied and assembled from a previously existing genome. The recipient *M. capricolum* provided a complete cell minus its chromosome as a recipient for the new DNA, with all its protein machinery in place in order to operate the new genomic instructions. This is where the computer analogy becomes quite relevant. But as a technical accomplishment in the field of genetic engineering, the paper ranks highly. Manipulating such large pieces of DNA is not easy, and the team had to generate new ways of manipulating a complete bacterial chromosome inside yeast. With time, the artificial synthesis of cells in the laboratory may well encompass a greater number of steps, including greater human design features in the genome rather than simply copying an existing genome. For those intrigued by the minimal number of components required to generate a living system, such experiments have some curiosity value. But otherwise, given that the living cells we already know about do such a great job, it's not that clear why pursuing the generation of evermore synthetic cells should be of great scientific interest.

Commercially the implications might well be different. At present, recombinant proteins for medical applications can be made rather efficiently using recombinant DNA technology, as we consider further below. The possibility in the future to construct totally new types of bacteria that could, for example, mop up oil spills without damage to the environment, or synthesize new energy sources, or make pharmaceuticals more efficiently than at present, is an alluring prospect. But whether generating quasisynthetic cells will achieve such aims more efficiently, only time will tell. Genetic engineering of existing bacteria and other types of cells has already taken us a long way toward some of these goals. Biotech companies have taken the first steps to engineer bacteria to make the kind of hydrocarbons that could be used as fuels, and vaccines are already routinely made using recombinant DNA technology. The U.S. Department of Energy has set a goal of replacing 30 percent of current petroleum use with fuels from renewable biological sources by 2030, and dozens of companies are involved in the race to help achieve that goal by the use of all kinds of genetically engineered organisms.

What about the risk of misuse? Ever since the advent of genetic engineering, there has been the risk that criminals or terrorists might abuse the techniques to create new and ever more pathological species of bacteria, or toxic compounds, with which to launch an attack or threaten a population. Such possibilities should be treated seriously, and the regular appearance of such scenarios in movies is at least useful in keeping such possibilities alive in the minds of government funding agencies. In reality, however, a whole range of nasty bugs already exists that could readily infect a population if used in a malevolent fashion, so such threats are nothing new. As far as making synthetic cells with evil intent is concerned, it is good to remember that the 2010 announcement by the Venter team was the culmination of work by a team of around twenty people working on the project for more than a decade at a cost of more than $40 million. You would need to be a very well-trained, dedicated, and well-funded terrorist group to exert that kind of effort for an evil goal that conventional means could more readily achieve.

Genetic Engineering and Medicine

Therapeutic Reagents

The main impact of genetic engineering in medicine so far has been on the generation of dozens of different therapeutic reagents that can now be made more efficiently and in purer forms in comparison with earlier methods. Originally the insulin used to treat diabetic patients was purified from the pancreas of animals from the slaughterhouse, such as pigs and cattle. The problem was that this insulin was not identical to that found in the human body; being recognized as slightly foreign during a lengthy period of ingestion, the body would start to mount an immune reaction against it by the production of antibodies that in turn attacked the insulin, reducing its efficacy. Now, as mentioned above, insulin is routinely made in yeast or bacteria by genetic engineering, and the product is indistinguishable from the insulin produced in the human body. One great advantage of making human proteins in bacteria is that they do not support the propagation of animal viruses, thereby reducing to zero the risk of infection.

Chapter 1 gave hemophilia as an example of a genetic disease that displays an X-linked pattern of inheritance. Patients used to be treated with the missing Factor VIII (required for blood clotting) following its purification from donated blood. The problem was the viral infections that came along with the regular blood donations, and as many as 90 percent of hemophiliac patients were infected with HIV or hepatitis C before the advent of genetic engineering enabled production of the pure protein. This provides an example of when bacteria are not the best cells in which to produce a protein. The reason is that many proteins, of which Factor VIII is an example, are modified by the addition of various sugar groups. In some cases, if these sugar groups are not present, then the protein does not function properly. Bacteria do not use the same sugar modifications for proteins as cells do in animals. In practice, then, Factor VIII is normally made at industrial levels in Chinese hamster ovary cells that have proven to be highly convenient for genetic manipulation and in their ability to be grown successfully in the large 500-liter bioreactors needed for production. These cells, unlike bacteria, have the ability to make the correct sugar modifications on the Factor VIII.

A huge market is also being tapped for the production of monoclonal antibodies, specialized proteins produced by human-mouse hybrid cells; the term "monoclonal" refers to the fact that the antibody is made in one clonal cell line and is highly specific for one specific foreign protein. Their mouse origins render them unsuitable in their original form for therapeutic use in humans due to their recognition by the immune system as foreign proteins. But by genetic engineering, monoclonal antibodies can now be "humanized" in such a way that they no longer look foreign to the human immune system, opening the door to a wide range of therapeutic uses, such as the specific targeting and destruction of cancer cells. Monoclonal antibodies are used in hundreds of clinical trials and in thousands of diagnostic tests. Their huge advantage lies in their specificity.

Dozens of other reagents made by recombinant DNA technologies continue to bring reassurance to patients that their medications are pure, well tested, and of the highest standard.[3] Those genetic pioneers who first started experimenting with DNA and restriction enzymes had little idea of the huge medical potential that their very basic biological research would realize.

Gene Therapy

In chapter 9 we introduced the use of gene therapies in the treatment of genetic diseases. One of the big challenges is to find suitable vectors to convey the missing gene to the right tissues without causing immune or other reactions in the patient. The modest successes already achieved using genetically modified stem cells were highlighted (in the cases of SCID due to adenosine deaminase deficiency). A similar approach has been used successfully in the treatment of several young boys in Paris suffering from the X-linked disease adrenoleukodystrophy (ALD). The disease affects about one in twenty thousand boys and develops during the age of six to eight years, leading nearly always to death in adolescence. The lethal neurodegenerative condition is caused by a mutation in the odd-sounding ABCD-1 gene which leads to abnormally high levels of a fatty acid that damages the insulation surrounding nerve cells. The 1992 tearjerker *Lorenzo's Oil* featured one family's struggle to save their son from ALD by the use of a dietary supplement, an oil that never fulfilled their hopes as an effective cure. But the reported gene therapy trial showed that stem cells expressing the ABCD–1 gene stopped the development of ALD in its tracks in two young boys, with further boys still being treated.[4]

Vectors that are showing promise in gene therapy trials include modified forms of various types of virus, such as the adenovirus and the lentivirus. In fact, a lentivirus was used to transfect the stem cells used in the ALD gene therapy just mentioned. Adenoviruses cause around 10 percent of the respiratory infections found in children and are a frequent cause of diarrhea, so of course they are genetically modified to disable such outcomes when being used as recombinant gene vectors. They have the advantage that they do not integrate their DNA into the genome of the host cell, so reducing any risk of causing cancer. They gain ready entry into cells via receptors on the cell surface and then duplicate themselves once inside the nucleus. In this way they can be used to bring a missing gene right into the cell, which then produces its protein product by transcription in the usual way. The therapy is temporary as the adenoviral DNA (with the replacement gene that it contains) is lost once the host cell replicates. Therefore the adenovirus is best used as a vector for a specific short-term task, such as the targeted killing of cancer cells. Hundreds of gene therapy clinical trials have been carried out using modified adenoviruses, as well as other types of viral vectors.

A wide range of newly discovered approaches are also under development for the genetic therapies of the future, increasing hope that the initial leaves on the gene therapy tree will soon lead to a full greening of all its branches, many of which still remain bare. For example, in chapter 2 we illustrated the way in which microRNAs can inhibit the actions of genes. Processing of certain microRNAs can result in the production of interfering RNA (or RNAi for short). This happens under natural conditions inside cells, and RNAi molecules can also be synthesized as very useful tools in research (used for what is often referred to in lab-speak as "gene knockdown"). The therapeutic potential of RNAi was recognized very soon after its discovery in the worm *C. elegans* in 1998.[5] One great advantage of these reagents, if well designed, is their potential selectivity in inhibiting the actions of a single gene. Since many diseases are caused by the inappropriate location or hyperactivation of one particular gene, the therapy goes straight to the heart of the problem: the actual disease-causing molecule. RNAi is so specific that it can distinguish between the disease-causing allele of a gene and its normal counterpart. If the mutant gene is the target, then this is potentially an important way to avoid side effects, as the normal gene function will be left untouched, whereas the actions of the disease gene will be blocked.

Clinical trials using RNAi are in progress for a wide range of conditions, including AIDS caused by the HIV retrovirus, cancers caused by oncogenes (the target genes), and macular degeneration of the eye in which a gene is targeted that encodes a growth factor necessary for the abnormal growth. Cellular genes required for HIV entry or replication have also been targeted because this avoids the problem of genetic variability in HIV: the retrovirus is a moving target, constantly mutating in each HIV-positive person, making it a nightmare to attack therapeutically. As with other genetic therapies, a major challenge is how to convey the RNAi reagent to the right place in the body. With the eye, the task is simple: the reagent can be injected directly. For other tissues, incorporation into aerosols, liposomes, and attachment to tiny nanoparticles are all being explored as the means of conveyance. RNAi therapies are in their infancy, but the potential for success is certainly there.

Genetic Modification of Animals

The first question that might come to mind is: why would anyone want to genetically modify animals? As far as biomedical research is concerned, the answer is: because this provides one of the most powerful approaches to understanding killer diseases like cancer; the diseases of the immune system, like arthritis, diabetes, and psoriasis; and neurological diseases.

The genetically modified mouse has been at the forefront of biomedical research now for more than two decades (though in 2010 the first genetically modified rats were reported as well). The question is often asked: why can the same kind of research not be carried out using cell lines cultured in the laboratory? Cell lines refer to animal, plant, or human cells from a particular tissue that are stimulated to grow in the laboratory under controlled conditions; cancer cell lines grow spontaneously without any need for stimulation, because that's precisely the property that causes cancer. Indeed, cell lines are incredibly useful—we could not do without them—but at the end of the day, the immune system (for example) is just that, a system. It consists of hundreds of different biological components that all have to operate together in order to generate the amazing system that defends us so well (most of the time) against the foreign invaders that seek to gain entry into our tissues: bacteria, viruses, and parasites. To understand the system, the whole animal context is vital, because that's where the system operates. A cell line can tell you much about how the individual cells of the immune system function, but only the whole animal can tell you how all those various cell types operate together to make up a system.

The same is true when investigating cancer. For sure the cancer cell lines are hugely useful in research, but cancers happen in bodies, with all this implies for initiation, growth, and metastasis. My own research group would never have been able to track down certain key mechanisms in cancer unless we had first noticed the quite unexpected finding that some of our mice were starting to die from a cancer very similar to human T-acute lymphoblastic leukemia, a finding that eventually led us from mice to cancer cells obtained from patients, and a potential new approach to therapy.[6]

The precise techniques for genetically modifying mice are available in any good genetics textbook and are not described here. Basically there are two approaches, either to overexpress the actions of a gene by making transgenic mice, or to remove a gene altogether by making "gene knockout mice," in either case making the genetic changes in the germ line so that a whole interbreeding colony of mice can be generated with exactly the same genotype. To make interpretation of results even clearer, the gene of interest can be deleted or overexpressed in one specific tissue, and even switched on or off at will, by the simple expedient of adding a chemical to the drinking water.

In practice, most of our knowledge of the role of specific genes in the functioning of the immune system, as in other research fields, has come from such transgenic and knockout experiments. The brief way the approach has been summarized makes it all sound very easy, but in practice, challenging research hurdles often occur along the way. Occasionally some long-suffering PhD student spends three years making a gene knockout colony of mice only to find (just when funding is running out) that the mice are fine and hunky-dory, with no apparent need for that gene at all, thank you very much. Life at the lab bench is not always easy, but the rewards are great for those who persevere.

Genetically modified animals are generated not just for research, but for medical purposes as well. For those suffering from a rare genetic condition called hereditary antithrombin deficiency, a group of goats living in Framingham, Massachusetts, might just be lifesavers. In 2008 the FDA gave the first permission in the United States for the genetic modification of an animal in order to produce a therapeutic reagent for use in humans. Antithrombin is an anticoagulant that prevents blood clots in normal human blood and has to be taken regularly by those suffering from the deficiency disease. These goats have been genetically modified to produce the antithrombin in their milk; all that needs to happen is for them to be milked in the usual way, and the antithrombin is then purified from the milk. Genetically modified sheep are likewise being milked for the protein alpha-1-antitrypsin (AAT) needed for the treatment of emphysema, an incurable disease involving degeneration of the lungs.

Pigs have been genetically engineered for rather different purposes. Some have been genetically engineered to produce human hemoglobin for blood transfusions. Other pigs have been genetically modified to remove certain antigens from their tissues. The antigens are proteins bearing sugars that are recognized very vigorously by the human immune system, so causing rejection of transplanted pig tissues. By removing the genes encoding the antigens from the pig genome, the hope is that pig hearts or kidneys, for example, could be used for transplant into human patients.[7] Before getting too strong on the "yuk" feelings that might (understandably) arise from having a pig heart beating in your chest, it's good to remember that if you're dying from terminal heart disease with no human organ donor in sight, then even a pig's heart might seem like quite a good option. In fact, plans for all these transgenically modified pigs came to a halt with the realization that we simply don't know enough about the possible transfer of pig viruses into the human population that might result from such transplants. So the genetically modified pigs, important as they are, wait on the back burner until the day arrives when such regulatory challenges have been addressed satisfactorily.

Genetic Modification of Plants

The genetic modification (GM) of plants has aroused probably more passions than any other type of genetic manipulation. The aim here is to be descriptive rather than proscriptive. Whatever one's personal views may be about the matter, it is good to know what's going on, and the aim of this section is to provide a brief overview, raising just a few of the complex environmental and social issues along the way.

Getting Foreign DNA into Plants

Getting foreign DNA into plants requires a somewhat modified tool kit in comparison with the process for animal cells. Plant cells have tough cell walls that are not so easy to penetrate. In the late 1980s, scientists came up with what the Brits might deem a "typically American" solution to the problem: shoot your way in—literally. DNA sticks well to very small gold or tungsten beads, and these could be shot into plant cells in such a way that, in some cells at least, the foreign DNA would become incorporated into their chromosomes. In the early models of this "high-tech" method, the barrel of a .22-caliber gun was sawed off and used for DNA blastoff. The trick was to get the gun close enough to the plant cells so that the DNA particles got right inside the cells, but not so close that the cells were blown to bits. Not a very natural method in this case, but it works. Fortunately DNA is a very robust kind of chemical (which is incidentally why some people think that life on earth might have started by DNA hitching a ride on a meteorite and arriving from some other planet). The kind of "gene guns" used today keep to the same idea, but the quick release of helium under pressure is now used to generate the explosive force.

A quite different method depends on the soil bacterium *Agrobacterium tumefaciens*, the causal agent in a plant disease called crown gall in which tumors form on the infected plant (yes, plants can get cancer as well). The bacterial plasmids contain genes that cause the tumors, and the bacteria have found ways of penetrating the plant cell walls; once inside, the plasmid DNA is incorporated into the genome of the infected cells. For genetic engineering purposes, the tumor-inducing genes are simply removed from the bacterial plasmid DNA and replaced with the foreign DNA that one wishes to insert into the plant cell genome. The whole process takes only a few hours.

Commercial Use of GM Plants

GM foods first started coming to the market in the early 1990s and for the most part have been transgenic plant products such as soybean, corn, canola, and cottonseed oil. The main modifications have been made with the aim of producing resistance to insects that destroy the crops, or to protect against certain types of herbicide so that weeds can be destroyed without affecting the food-producing plant.

The manipulation of plants so that they contain their own insecticide came with the isolation of toxin from the bacterium *Bacillus thuringiensis* (Bt for short). Different strains of Bt (there are thousands) produce toxins specific to different orders of insect, and some are lethal to the pests that attack crops such as cotton. The Bt toxin is a protein that breaks down quickly in the environment and is nontoxic to humans and animals, so is much "cleaner" in its impact than spraying fields with insecticides—in the process also saving huge amounts of water. Using the *Agrobacterium* technique, the Bt toxin has been introduced into a wide range of crops, including cotton, tomatoes, corn, and potatoes. One problem is that Bt is not toxic to all pests in a given agricultural environment, so destruction of one type of pest can allow others more scope for proliferation by a type of natural selection.

Plant resistance to glyphosate and glufosinate herbicides has also been a major development in the genetic engineering of crops. The gene that bestows resistance has been incorporated into soybeans, corn, rapeseed, and sugar cane, among others. Herbicides such as glyphosate are very efficient at clearing weeds and they are significantly less toxic than the previous generation of herbicides, though not as safe as Bt toxins.

Three concerns have arisen, however, from the widespread use of these GM crops. The first arose from a systematic trial carried out in the United Kingdom involving two hundred test plots in which various herbicide-resistant GM plants were grown alongside conventional crops. The destruction of weeds was so great that it led to a significant loss of biodiversity—for example, 24 percent fewer butterflies in some sites. Of course, loss of biodiversity can occur also with conventional herbicides: this is not unique to the GM situation. The second concern is that herbicide-resistant genes might spread to weeds, meaning that herbicides such as glyphosate no longer kill the weeds. This does not appear to be a significant problem, although in some reported cases the gene has been detected in nearby non-GM crops.[8] However, the third point, that weed resistance to any herbicide can occur by natural selection, is certainly a problem with glyphosate simply because of its massive use. But again this is not a glitch unique to GM, although introducing a completely re-engineered crop is clearly more complicated than simply switching weed-killer in the case of a non-GM crop.

In addition to pest and herbicide resistance, plants have also been genetically modified with the aim of increasing their nutritional value. "Golden rice" incorporates three new genes, two from daffodils and one from a bacterium, and is rich in a precursor of Vitamin A that the body can readily convert to this essential nutrient. Various strains have been produced, and the goal is to deliver the necessary daily dose of Vitamin A in 100 to 200 grams of rice, which is the average daily intake by children in rice-based populations, such as India, Vietnam, and Bangladesh. The WHO estimates that 250,000 to 500,000 children every year become blind due to Vitamin A deficiency. More than half the children who lose their sight die within one year, often due to immune deficiencies that are also associated with a lack of Vitamin A. Yet the use of golden rice has been delayed since it was first developed in the laboratory in 1999 by the weight of regulatory requirements.[9] The introduction of such GM foods needs to be viewed in the context of a wider program to improve the nutritional health of children, and perhaps a more obvious solution to the problem than GM foods is the use of a more varied diet. Green vegetables and unpolished rice are good sources of Vitamin A. Nutritionally supplemented foods are no magic bullet, but one important contribution among many to the challenge of providing a healthy diet.

Plants have also been genetically modified for quite a different purpose: to generate vaccines in their cells so that people might be immunized against pathogens even as they eat their breakfasts. Although still in the development phase, projects have shown considerable success under experimental conditions. For example, bananas have been genetically modified to produce the human vaccine against Hepatitis B.[10] Plants provide safe and economical systems for the production of human proteins. Chloroplasts contain their own small amount of DNA, just as mitochondria do in animal cells. There can often be more than ten thousand copies of the chloroplast genome in each plant cell, so incorporation of a foreign gene into the chloroplast genome means that any gene expressed will have a flying start. Indeed, nearly half the leaf protein can consist of the product of a foreign gene incorporated into chloroplasts. For example, one acre of chloroplast transgenic plants can produce up to 360 million doses of clean, safe, and fully functional anthrax vaccine antigen.[11] The vaccines of the future will most likely be generated by growing a few acres of a transgenic crop every now and again; the harvest should be sufficient to keep a whole country going for some time.

GM Plants and the Challenge of Food Production

Increasing the productivity of land for food production provides a key for feeding the hungry mouths of the twenty-first century, and genetic engineering will play its part in this huge challenge. World population growth is expected to start reaching a plateau by 2050 at around 9 to 10 billion people, one-third more mouths to feed than the present 7 billion. No one doubts that the planet has the capacity to feed that many people, but at what cost? The "green revolution" that saved the world from mass starvation in the latter half of the twentieth century depleted the soil of water and contaminated the rivers with fertilizers. Only about half the world's land that could be farmed is currently being farmed, and the remaining 50 percent, about 1.6 billion hectares, is largely to be found within Africa and South America.[12] But, as Britain's Royal Society has suggested, a far better solution than expanding farming areas involves what has been termed the "sustainable intensification" of present farmlands in order to increase yields. This approach will involve a whole range of changes, of which genetic engineering can be seen as one component.

One of the biggest challenges to the plant genetic engineers is to modulate crops in such a way that they become more drought-

and heat-resistant. The stresses that affect plants include "drought, salinity, heat, cold, chilling, freezing, nutrient, high light intensity, ozone and anaerobic stresses."[13] U.S. farmers lose on average 10 to 15 percent of their annual yield due to drought and water stress. Making life for plants (and for molecular biologists) even trickier is the fact that the effects of two stresses are not simply additive, but synergistic. A comparison of all major U.S. weather disasters between 1980 and 2004 showed that a combination of drought and heat stress caused more than $120 billion worth of crop damage, whereas drought not accompanied by heat stress resulted in "only" $20 billion worth of damage. The two together proved the real killer.

Fortunately a new wave of stress-resistant plants is already under development, and drought-tolerant biotech maize is expected to be ready soon for commercialization. This provides yet another example where "borrowing" a gene from a bacteria has a beneficial outcome. The gene in this case is known as *cspB* and it comes from *Bacillus subtilis,* in which it helps bugs adapt to stress such as that caused by a change in temperature. The same gene helps maize to cope with drought by reversing the entanglement of RNA that tends to occur under drought conditions. The idea is that the maize spends a lot of energy in repairing its RNA, and by speeding up this process *cspB* enables the energy to be used in helping the maize to grow well even in a drought.

Improving plant roots is also a key strategy in the second wave of new food crops that, it is hoped, will keep the world population alive during the twenty-first century.[14] Currently only 40 to 50 percent of the nitrogen from fertilizers gets into plants, so designer roots are envisaged that will be far more efficient at taking up nitrogen. Ideally plants could be designed that would fix nitrogen directly from the air in the same way that they utilize carbon dioxide, but all attempts so far in this direction have failed. An underground revolution in root design seems the approach most likely to succeed for the moment. Food crops with longer roots that will reach down farther into the soil to extract water and nutrients are also being developed, both by traditional breeding methods and by genetic engineering. The bread wheat genome has a huge 17 billion base-pairs, whereas a grass in the same subfamily as wheat called *Brachypodium distachyon* has a genome with a more manageable 271 million base-pairs, so a project is under way to screen this smaller genome for genes that increase root length that could then be incorporated into bread wheat. Other projects aim to isolate roots that are more resistant to root rot caused by a fungus. All of these developments are part of the sustainable intensification that, it is hoped, will help feed the world in the coming decades without massively increasing farmland.

Use of GM Plants Worldwide

North America has most enthusiastically embraced the use of GM crops, although many other areas of the world are catching up fast. In the United States, more than 90 percent of soybean crops are GM, as well as around 80 percent of the cotton and 70 percent of the corn.[15] As far as the rest of the world is concerned, it took ten years for the first billion acres to be planted with GM crops, up to 2005, but only three years, up to 2008, for the second billion.[16] More than 13 million large, small, and resource-poor farmers in twenty-five different countries are planting biotech crops. Significantly, 90 percent are in the small and resource-poor category. China, India, Argentina, Brazil, and South Africa, with a combined population of 2.6 billion, have been leading the way internationally. Brazil now has twenty-one types of GM plant approved for use and is second only to the United States in the number of hectares planted with GM crops.[17] GM soya already makes up more than 75 percent of the Brazilian soya market. Countries with large populations are doing all they can to seek an increase in yields and to lower the cost and environmental impact of pesticides.

Perhaps the least enthusiastic response to GM foods has been in Europe, although even there some Bt maize is planted and

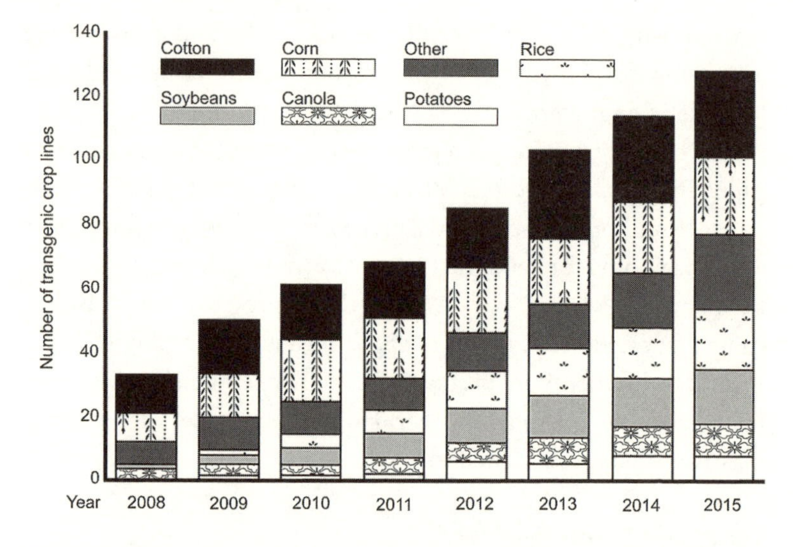

FIGURE 11.5. Current numbers and estimations of future numbers of genetically modified (GM) crops worldwide. Although the commercialization of the crops shown may be technically possible by 2015, the regulatory feasibility may be more questionable given that in some countries no GM (food) crops have been authorized so far. Adapted by permission from Macmillan Publishers Ltd. Stein, A., and Rodríguez-Cerezo, E. *Nature Biotechnology* 28 (2010): 23–25.

attitudes are not uniformly negative. But European Union politicians in twelve years have approved just two varieties of GM crops for growing, compared to more than 150 worldwide during the same period. (Figure 11.5. Note that the number of crops being grown, as the figure illustrates, is less than the number that has been approved.) Commercial GM planting in Europe in 2009 covered less than one hundred thousand hectares, mostly in Spain, compared to 134 million hectares globally. Why the reluctance? One reason is simply the fact that Europe has plenty enough food production already. Farming subsidies are widespread in order to keep countryside under cultivation and to keep farmers in business who sell at a loss because of overproduction. But there is also a powerful Green Movement that often opposes GM crops on the grounds of supposed safety concerns or the worry that they are not "natural." Underlying all this are suspicions about the role of big multinational agrifood companies; about the public statements of scientists, coupled to general antiscience attitudes; and about cases (as in the infection of cattle by BSE in Britain) when government reassurances about food safety were not born out by the actual outcome. The abuse of science in general, and of genetics in particular, in Nazi Germany has also left in its historical wake a long-term mistrust of genetic manipulation.

One of the practical outcomes of European resistance to GM foods has been barriers to imports from African countries that would like to export GM foods to Europe, but are prevented from doing so by European legislation. This in turn has inhibited the planting of GM crops in African countries, leading to the view that Africa has been starved of technology.[18]

Opposition to GM foods purely on ideological grounds—for example, based on the claim that it is "not natural"—seems hard to sustain. The huge amount of gene transfer that has occurred throughout the evolutionary bush of life reminds us that no living thing is static. We have already seen how full our own bodies are of ancient genes that also occupy the genomes of bacteria, yeast, bananas, coral, fish, and kangaroos. So the best approach seems to be to address each case for GM on its own merits. Does it really have practical or commercial advantages? Have the safety issues been adequately assessed? Will its introduction benefit the poor? What effect will it have on biodiversity? And what impact might it have on traditional farming practices and local community structures? Farmers have traditionally stored up seed from this year's crop to plant next year, a practice that goes back millennia, although evidence suggests that keeping seed year after year leads to a decline in its quality due to viral infections. But with certain GM crops, farmers have to sign up to purchase a new batch of genetically modified seed each year from a commercial company. Something of the farmers' freedom and autonomy is being lost here, although several companies are now making the seed freely available due to public pressure. Certainly there are insufficient social studies on the various ways in which GM crop planting may have deleterious effects on local community life. What are the relational costs of introducing GM crops to a certain community? These questions are not all necessarily that easy to answer. But asking the right questions is important in conducting public discussions in which extreme, polarized views can so readily dominate the debate.

In this chapter we have said much about the genetic manipulation of plants and animals, and about their use for medicine and other aspects of human welfare. But what about the genetic manipulation of humans? Do we somehow belong in a different category from plants and animals, such that different principles apply to us? In the next chapter we begin to unravel this and some of the other big questions that the language of genetics raises with respect to our own sense of human worth and identity.

CHAPTER 12

Genes and the Big Questions of Life

THOUGH GENETICS IS a science, it also forces us to look beyond science.[1] It raises questions about our identities and our role as caretakers for our DNA and that of other species. Bears, kangaroos, and wombats don't care what happens to each other's DNA, nor do leopards, beetles, and woodpeckers. But we do. Uniquely of all species we have the conscious awareness of what is going on in the world, and with that comes all kinds of responsibilities. Some people see genetics as the savior of humanity. Others see it as a threat, dehumanizing or giving humanity powers not matched by human wisdom. In this chapter we consider some of genetics' wider questions that lie beyond science. These questions, while not scientific, can be considered just as rationally and carefully as any scientific question.

GENETICS AND HUMAN IDENTITY

The challenge and insight that genetics brings to our sense of individual worth and identity arises from its power to unravel the very language of life. Surely, so one argument goes, it would be better to leave life as a mystery. Are there not just some things that would be better not to know? Where is human identity amid all the molecules? Such concerns may be exacerbated by comments such as these from the biologist Richard Dawkins, as he assured a lecture hall full of schoolchildren that

> We are machines built by DNA whose purpose is to make more copies of the same DNA. . . . That is EXACTLY what we are for. We are machines for propagating DNA, and the propagation of DNA is a self-sustaining process. It is every living objects' sole reason for living.[2]

If one's sole reason for living is to exist as a "survival machine" to pass on one's genes to the next generation, that doesn't sound like much fun. Indeed, taking a long view, if that's the sole reason for living, then it seems a bit pointless. Furthermore, if we are "nothing but" one out of billions of other survival machines, then the idea of human identity seems to get lost in all that machinery.

Fortunately, contemporary genetics continues to highlight the uniqueness of each individual. Machines rolling off the same factory line are identical, but we are not. At the DNA sequence level, as we have seen, the variation between individuals is far greater than envisaged even just a few years ago. As thousands of human genomes are sequenced, the extent of this interindividual variation will be further refined at both the genetic and epigenetic levels. Not even identical twins are really identical.

Each human life is a story about a unique genome tightly interwoven with a unique environment generating a unique individual. The interaction between genes and environment begins at the earliest stage of fetal development—or even further back in the parents and grandparents—and continues throughout life. There is never a moment when genes and environment are out of contact, mediated via the human body itself. And people are constantly making choices that, as we now know, make an impact upon their own genomes at the epigenetic level. Choose smoking or overindulgence in food or alcohol, and you choose to change your epigenome, possibly with implications for your progeny as yet unborn.

Consider this thought experiment: What would happen if all the information content of a single individual's brain were transferred to another human body from which the information (thoughts, memories, etc.) had been deleted? Would that body then represent the very same person as the original? In reality, within less than a second the "new" person would be different from the information-donor, because neither his environment nor his decisions would be exactly the same, and divergence into a quite different personality would rapidly occur as his brain state began to vary. An analogous point applies at the genetic level as well, albeit on a slightly slower time course: if an individual's genome were transferred to another individual whose own genome had been (magically) deleted, then epigenetically that person would become different within a very short space of time, especially considering that epigenetic modifications are unique for all the two hundred or so tissues of the human body. Genetics provides a very solid basis indeed for the assertion that every human individual is truly unique.

The dynamic complexity of the human body, interwoven with the environment, is also highly relevant in the light of claims about genetic determinism. The headlines typically report discovery of a gene "for" some particular trait, be it athletic prowess, musical ability, sexual orientation, criminality, intelligence, or alcoholism. But as we have seen, there are no such genes "for" complex human traits. Thousands of genes are involved. Human individuals decide what to do with their lives. For sure, their genetic endowment may on occasion restrict their choices: the person five foot tall may find it more difficult to become a top basketball player than someone who is six-foot-five. But plenty of people around who are six-foot-five choose not to play basketball, and some who are five feet tall do. Our genes may sometimes shape the range of our choices, but they do not determine them.

The way that we view human responsibility is well illustrated by the case of the Y chromosome. A very striking correlation is present between possession of the Y chromosome and levels of criminality in every country of the world. Out of 131 countries worldwide, an average of 96 percent of the prisoners are male.[3] So universal is the correlation between the Y chromosome and criminality that we can safely say that no other genetic correlation will ever be found between a variant genome and criminality that surpasses this one. And yet we still hold nearly all males responsible for their criminal actions and put them in jail as soon as they're convicted. Furthermore, we note that most people who possess a Y chromosome go through life without committing a crime. So having a Y chromosome, with its unique set of genes, does not "determine" human criminality, although clearly we cannot go to the opposite extreme and say that it has nothing to do with human behavior at all.

The human genome endows us with freedom of choice. It gives us brains and nervous systems in bodies of sufficient sophistication and complexity that we are able to weigh up the consequences of our actions. Brains don't make choices, but thinking, reflecting people do (and they certainly need a normally functioning brain to exercise their free will!). Entire books have been written on free will, but suffice it to say that nothing in genetics counters such a notion.

If genetics gives us no reason to deny that people are responsible for their choices, it also points out the rare exceptions. In cases of significant genetic abnormality, mental development can be thwarted; individuals may not know the consequences of their actions. This is recognized in the legal system, and psychiatrists have the difficult task of testifying in court as to when this is the case. For example, the court may consider whether other people who carry the same genetic or chromosomal variant also commit crimes, a kind of analysis that is not decisive, but relevant in such medical cases.

In the future, defense lawyers will doubtless use more arguments about genetics and brain function in criminal cases, claiming that the client is not accountable for his or her actions. But the example of the Y chromosome should always be kept in mind. Treating people as if they are mentally sick, when in reality they are not, represents a diminishment of their humanity. The default position is that we should treat people as responsible human beings unless we have very good reason to the contrary.

Outside the courtroom, however, we are left with a somewhat more philosophical challenge as to how to relate the genetic descriptions that have dominated the previous chapters to narratives about human free will, morality, and purpose. Perhaps the most helpful approach is to see these different narratives as complementary levels of understanding: we need them all to do justice to the complex reality of being a human being, and there is no particular reason why any of the levels need be in conflict with each other.

The molecular genetic level is the one that we have been considering through this book. We have seen how genes build cells. Cells have "emergent properties," that is, properties that are not reducible into the language and concepts of genetics. Describing the body at a cellular level requires the language of cell membranes, nuclei, mitochondria, signaling pathways, and metabolism. Millions of cells together in turn make up different tissues that also have emergent functions that cannot be contained within the language and concepts of the cell. For example, the heart acts as a four-chamber pump that keeps the blood oxygenated and flowing around all the tissues of the body. The notion of "being a blood pump" is not found at the cellular level. It is the property of many cells arranged in a certain way.

In turn, whole bodies also have unique emergent properties that are nonreducible into the language and concepts of individual organs like hearts and kidneys. They run, walk, eat, have sex, and do many other things that require a whole new level of explanation using the language of agency. The notion of "mind" in turn relates to yet another set of explanations that involve conscious experience, "mind" being an emergent property of brain. However complete our descriptions may be of the functioning of the 10^{11} neurons that constitute the brain, the notion of "I need to go to the store and buy some bread" is never going to be encompassed within such neuronal descriptions. The "it" descriptions provided by the neuroscientist of my brain activity are complementary to the "I" descriptions of my self-conscious reflection.

In this approach, we are not talking about anything "mystical" or "spooky" moving the body, as in the *élan vital* once proposed in the past. Instead we are using different complementary narratives to describe life's complexity. Together they make up a single reality. And since no one is up to the herculean task of comprehending that reality at one go, in practice we must take one narrative level at a time. As a geneticist I would be the first to recognize that there are overenthusiasts in my field who claim genetics is the *only* narrative that counts. This can happen in any field, of course. Genetics provides us with good explanations for the biological properties of organisms, but it is hardly the case that human life is "nothing but" genes in action.

This kind of "nothing-buttery" has a kinship with scientism. We saw earlier how Dawkins sailed close to "nothing-buttery" by saying DNA replication is the "sole" reason for living. Similarly James Watson, codiscoverer of the double-helical structure of DNA, offered these immortal words: "There are only atoms. Everything else is merely social work."[4] His fellow discoverer, Francis Crick, who moved on into brain research during the later decades of his life, once commented, "The Astonishing Hypothesis is that 'You,' your joys and your sorrows, your memories and your ambitions, your sense of personal identity and free will, are in fact *no more than* the behavior of a vast assembly of nerve cells and their associated molecules."[5] But it's not clear why primacy should be given to the scientific description of "nerve cells and their associated molecules" rather than to the "joys and sorrows," which, after all, represent the primary data of our conscious experience. Likewise there is no need to think that descriptions of genes and DNA are in any sense more "real" or "valid" than the experiences of the conscious human being to which they give rise. As we have already considered, a description of the human person simply at the level of the genes is quite insufficient to explain conscious experience.

Such reflections are relevant to those with religious beliefs who wish to claim that human existence has some kind of overall meaning and purpose. In the Judeo-Christian tradition, in particular, there is the claim that every human being is made "in the image of God,"[6] meaning, among other things, that divine authority has been delegated to humankind to care for God's earth. The claim also entails the value of each human being, independent of his or her genetic endowment, ethnicity, language, intelligence, or any other variant quality, resting on the fact that he or she is loved by God and entrusted with the responsibility of being his earth keeper. Being made in the "image of God" also involves endowment with a suite of capabilities that render relationship with God a possibility: big frontal lobes in the brain that facilitate conscious experience and moral decision making, linguistic abilities, a "theory of mind" that facilitates a life lived in relationship, and brain structures that allow conscious and rational reflection on the universe that God has brought into being.

Here we see theology as another level of narrative, complementary to other levels already cited. Theological accounts are not in any kind of rivalry with the scientific accounts, which are fine as far as they go. The insights provided by human genetics, of beings who are exquisitely constructed to be uniquely different from every other human being who has ever lived, or is likely to live, on this planet, are perfectly consistent with the notion of a God who loves each person as a human individual, endowed with all those capabilities required to render a relationship with God a possibility. Those capabilities may be generated by the information content of our genomes, but to reduce them to the level of the gene is simply a category mistake, as if all the functionalities of a car were conflated with the detailed design plans according to which they were constructed.

Those who understand humankind as being made "in God's image" see this as the most fundamental of all narratives that underlie human value and identity. The narrative provided by genetics is consistent with such an understanding.

Genetics, God, and Evolution

Traditionally many in the Abrahamic faiths have seen an evolutionary process as God's chosen means of bringing about biological diversity on this planet. Although we should be careful not to read back into historical texts modern understandings of the term "evolution," many ancient religious texts point to the general idea that God has created organisms through a lengthy process.

Ibn Khaldun, the fourteenth-century North African Muslim philosopher, historian, and polymath, exhorted his readers, "Look at the world of creation, how it started from minerals, then plants, then animals, in a beautiful way of gradation and connection . . . where the meaning of connection in these creatures is that the end of the horizon for each is ready in a strange way to become the first in the line of what comes after it."[7] Underlying such reflections is the ancient notion of the whole of life as representing a "Great Chain of Being" in which all living things are arranged in an orderly progression, not connected by common descent as in the Darwinism that came later, but certainly arranged in a logical order.

A similar theme is reflected in this remarkable passage from John Wesley, Anglican founder of the Methodist movement, writing a century before Darwin in his five-volume work, *A Survey of the Wisdom of God in Creation*:

> There are no sudden changes in nature; all is gradual and elegantly varied. There is no being which has not either above or beneath it some that resemble it in certain characters, and differ from it in others. . . . From a plant to man . . . the transition from one species to another is almost insensible. . . . The ape is this rough draft of a man; an imperfect representation which nevertheless bears a resemblance to him, and is the last creature that serves to display the admirable progression of the works

> of God! There is a prodigious number of continued links between the most perfect man and the ape.

The point in citing such authors is not to suggest that they were some kind of crypto-evolutionists, but rather to point out that theological resources were already available for the absorption of evolution into a theistic framework by the time Darwin's *On the Origin of Species* broke upon the scene in 1859. These thinkers of the Abrahamic faiths saw every aspect of God's creation, without exception, as reflecting the outworking of his actions in the created order. It was Darwin's brilliance to bring history into biology, and to show how the Great Chain of Being was in fact a historical tree in which every branch and twig were connected.

In contrast to the mythological accounts that have the church waging war on evolution in the years immediately following 1859, Darwin's new theory was in fact rather rapidly baptized into the traditional Christian doctrine of creation. Some theologians actively welcomed Darwin's theory of natural selection. They felt that it restored a much more active role to God's providence in creation when contrasted with the rather dry and distant notion of God promoted by the natural theology that had dominated during the first half of the nineteenth century. By the mid-1860s questions that assumed the truth of Darwin's theory were already appearing in the undergraduate exam papers of Cambridge University, that bastion of Anglican respectability where Darwin himself had earlier studied divinity. The 1860s also saw the first use of the term "Christian Darwinian."

But as with any totally novel new theory, Darwin's theory of natural selection did also receive some initial opposition. There were scientists who opposed the theory on scientific grounds alone or on religious grounds alone, or sometimes on both. Darwin's old mentor and teacher at Cambridge, the Revd. Adam Sedgwick—professor of geology, the one who had established the geological column, who had highlighted a very old age for the earth, and who had initiated Darwin into geology when a student by taking him on a field trip to Wales—opposed natural selection, thinking that Darwin's theory would undermine the unique value of humanity and subvert the moral order.

Often you find people today reading back into the 1860s the beliefs of today's young earth creationists as if that were the belief system of the time, but it really wasn't. Creationism of this type is a twentieth-century phenomenon. If people had an issue with Darwinism in the 1860s and 1870s, it wasn't generally to do with biblical literalism, nor with the age of the earth—everyone knew by then that the earth was very old—it was much more to do with concerns about human identity and the moral and political order.

Considering the present opposition to evolution in the United States, it's ironic that evolution was popularized in North America largely by Christian academics, such that the American historian George Marsden can write that "with the exception of Harvard's Louis Agassiz, virtually every American Protestant zoologist and botanist accepted some form of evolution by the early 1870s." In the words of the British historian James Moore, author of the definitive book tracing the reception of Darwinism in Britain and America in the nineteenth century, "With but few exceptions the leading Christian thinkers in Great Britain and America came to terms quite readily with Darwinism and evolution."

The theological resources available to those at the time of Darwin were rich indeed, stretching back through the Italian Dominican friar St. Thomas Aquinas (c. 1224–1274), then further back to the early church fathers such as Augustine of Hippo (354–430), and back even further to the texts of the Old and New Testaments. And this long tradition maintained a notion of "creation" that is distinctly different from that often envisaged in contemporary discourse. When we think about a person "creating" something, then we visualize an artisan or engineer bringing into being by deliberate intent every single aspect of their creation, so that its every detail reflects the mind and specific intentionality of the creator. But this is not what Christians mean when they speak of God as creator. The notion of "creation" in Christian theology refers more to ontology, the question of existence, why something exists rather than nothing.

Within this theistic framework, there is one great dualism: that between God and everything else that exists. All that exists only does so because God wills it into existence. The "creation" refers to that which God has willed into existence, and it goes on existing only as long as God wills that it does so. So the universe, or the multiverse—for the sake of this point it makes no difference—is itself the creation, with all its particular properties, its mathematical elegance, and its inherent intelligibility. Augustine was very clear that when God created, he brought into being the whole space-time continuum. Time itself is part of God's creation, and God himself is not bound by time, though he may readily engage with time-bound events.

The intrinsic intelligibility and rationality of the space-time continuum makes science possible. The properties of the matter and energy that make up the universe in this theistic view are what they are because God not only wills the universe to have these properties, but also actively sustains and upholds the universe in such a way that these properties are consistently reproducible. This is known as God's "immanence" in the created order. If God's immanence were not a continual reality, then science would be impossible and living in this universe would lack any coherence. Indeed, life of any kind would be impossible because life requires the consistency of the properties of matter so that it behaves in some ways and not in other ways. So in answer to the question "What difference does it make to your science if you believe in God?" the theist might have several different answers, but among them must surely be the most important answer: "Without God there wouldn't be anything in existence anyway, so certainly no humans and no science." Augustine had a very succinct way of putting the point as he wrote in the early part of the fifth century: "Nature is what God does."

Once we grasp the idea that nature is "what God does," then it is easier to see why theists should have no problem with the theory of evolution. If God has chosen to bring about biological diversity through a long historical process over billions of years, then so what? The task of scientists is to explore what God has done and continues to do in the created order, not to rationalize from first principles (as the ancient Greeks tended to do) what God ought to have done. There are no grounds for thinking that God "ought" to have done things in one way rather than the other. Of course, we are here not considering God as like the human artist or engineer who intentionally brings into being every detail of the created order, but rather seeing God, in the language of Aquinas, as the "primary cause" of all that exists, that existence being mediated by all kinds of "secondary causes" in the material world, the causes that are investigated by scientists. There is only one way of finding out what God has actually done, and continues to do, in the created order, and that is to gather data and test hypotheses experimentally in the way that scientists do. For the theist who also happens to be a scientist, all of his or her research is one long exploration of God's universe.[8]

This also explains why those who hold to this traditional understanding of creation are generally critical of the notion of Intelligent Design (ID), the idea that there are certain biological entities that are so complex that they could not possibly have come into being by the process of genetic variation and natural selection, thereby displaying the property of "design" and therefore, by inference, the existence of a "designer." It is difficult to see how such a proposal could be experimentally testable, so ID does not appear to belong to science. In the present context the main objection to ID is theological: it conjures up the idea of a heavenly engineer who occasionally tinkers around with the created order to produce some "design features" which stand out in contrast to the "nondesigned features" that, it is suggested, make up the rest of the created order. This is very distant from the notion of "creation" in traditional theism, which sees all that exists without exception as being God's creation. "Design" is a slippery word with many meanings, and Christians would certainly see the whole created order as being "designed" in the sense of fulfilling God's intentions and purposes. But this is very different from perceiving God as intentionally "designing" specific features of living things, an idea that forms no part of Christian theology (the idea of God as designer in this latter sense is, for example, found nowhere in the Bible).

In the context of genetics, three objections are often leveled against evolution: the first is that it seems a very wasteful process, the second is that it depends on "chance" in a way that seems inimical to the idea of a purposeful God, and the third is that it seems a very cruel process to use to fulfill the purposes of a God of love. The first two of these objections are easy to address, and we shall do so here, whereas the third objection is more difficult, and deserves a section of its own.

Is Evolution Wasteful?

On the question of waste, the staggering size of the universe with its 10^{11} galaxies each containing about 10^{11} stars provides a useful background for thinking about the subject. It has sometimes been suggested that the Creator is wasteful since he has made a universe which is so vast and so old. But the fact is that the universe needs to be this vast and this old in order for elements such as carbon and oxygen to be synthesized, and so for life to be able to emerge. The present size of the universe is related to its present age multiplied by the speed of light. If the universe were the size of our solar system, then it would last for only about one hour, clearly insufficient time for a fruitful earth! The universe has to be this big in order for us to exist.

The time factor is also vital in the emergence of all the biological diversity that we see on our planet today. Genetic change and speciation are relatively slow processes, and the Cambrian explosion happened only relatively recently in earth's history. Why then and not earlier? Further research may well clarify why this should be the case. The gradual increase in the oxygen level in the earth's atmosphere, once photosynthesis got going, is important in the diversification of life on earth, but many factors are involved. As our scientific knowledge increases, it might well be possible to start formulating more generalized "principles of emergence" that describe and predict the pace of evolution. The discovery of life on other planets would help in this process considerably. It is much more difficult to make generalizations when we only have one example, the life that exists on planet Earth.

Theologically it is difficult to know what "waste" might mean for God. Waste compared to what? Christians worship the God who says that "every animal of the forest is mine, and the cattle on a thousand hills. I know every bird in the mountains, and the creatures of the field are mine" (Psalm 50:10–11). The God who flings 10^{22} stars into space and "calls them each by name" (Isaiah 40:26) is also the great naturalist who enjoys all the richness and diversity of the natural world that he has brought into being, including its impressive carnivores like lions (Psalm 104:21). It is therefore not surprising that being made in God's image involves the delegation to some extent of the tasks of the heavenly naturalist (Genesis 2:19–20). "Waste" may be a relevant notion if you are running a factory or a business and need to be exerting tight control on expenditure, but it's difficult to see how the notion has any relevance when considering the fascinating history of evolution on planet Earth.

Is Evolution a Chance Process?

The idea that evolution is based on a chance process is based on a simple misunderstanding. One of Richard Dawkins' aims in writing his book *The Blind Watchmaker*, as he states in the preface, was "to destroy this eagerly believed myth that Darwinism is a theory of 'chance.'" It is indeed a myth, and the reasons should be apparent from the descriptions in previous chapters.

For sure the variation that comes into the genome by all the mechanisms previously discussed happens without the good or ill of the organism in view, nor can its precise location or extent be predicted, although we have noted that there are hot spots in the genome that are much more prone to mutations. But the variant phenotypes that it generates all have to be tested out by natural selection in the workshop of life. As noted in chapter 6, in the constant evolutionary interplay between "chance" and "necessity," necessity has the final word.

We also noted in chapter 6 the stringent constraints upon the evolutionary process as illustrated by the striking phenomenon of convergence, the evolution of proteins through a limited series of steps in "design space," and the notion of "fitness landscapes." In reality, evolution taken as a whole is a highly organized process. The evolutionary search engine is constantly seeking out the best adaptations for life in a given ecological niche. There are only certain kinds of carbon-based life that can exist on a planet with this particular range of environments, defined by a certain level of gravity, by hot to cold, wet to dry, high to low, above ground–below ground, and all the other myriad variations that define the homes for different species. The resulting millions of species are wonderfully varied indeed, but many of them are phenotypically very similar, and their shapes, sizes, means of locomotion, food preferences, and so forth are all highly organized consequences of their evolutionary histories. Evolution is no chance process taken overall, but a constrained series of events that have, to some extent, predictable outcomes. Evolutionary history does not force any particular metaphysical beliefs upon anyone (metaphysical beliefs are those that go beyond science). But for those who do believe in God, the evolutionary narrative is rather consistent with the idea of a God who has intentions and purposes for the world. The emergence of intelligent personhood from the process is what you might expect, given the existence of a personal God who wishes freely chosen relationship with those creatures made in his image.

Genetics and Life's Suffering

A significant challenge to the belief that evolutionary history arises from the intentions of a loving God stems from the huge amount of predation, pain, and death that have been involved in the process. Of course, for those who do not believe in God, or at any rate not a God of love, then this is not a problem—at least not this kind of problem. The universe just is, and its properties are just the way they are, and there is neither rhyme nor reason why they should be this way rather than another way. This stance illustrates the point that in the realm of metaphysics everyone has to start somewhere with some basic assumption and then move on from there. But scientists are trained to keep asking "why" questions. Why are things this way and not another way? Many find unsatisfactory the claim that there is no rhyme nor reason to the way things are, for the claim acts like a research blocker, hindering further metaphysical enquiry. By contrast, theists believe that the existence of a personal, creator God provides the inference to the best explanation for the intelligibility and rationality of the universe, and for the emergence of conscious, mindful, rational beings who can appreciate and understand its properties. So "the way things are" is no accident, but represents part of a larger plan of which we ourselves constitute one component.

Two kinds of evil are usually distinguished in this discussion—the "moral evil" that arises from freely chosen human actions, and the "natural evils" that arise from the intrinsic properties of the world: pain, predation, tsunamis, earthquakes, and all the rest. This second category of "evils" is often cited as problematic for those who believe in a God of love.

Many large books have been written on this question, and we

can certainly not solve the problem in a few short pages.[9] Only some brief reflections are offered here, in particular those that arise from the study of genetics. First, it might be worth querying the term "natural evil" itself. Everyone recognizes the moral evil that is committed by free human choice, but using the word "evil" about hurricanes and genetic diseases is clearly something rather different. Although the term is well embedded in the theological and philosophical literature about the subject, it does seem a rather odd use of the word "evil" when we compare it to the way we normally use the word. And when we look at it more closely, it's clear that what we're really talking about here is "things that we would much rather do without." Of course, we would much prefer that there was no pain or hurricanes or genetic diseases. But it's not clear that because we don't want these things, we should therefore label them as "evil." After all, it's not as if any of these things had any choice about the matter.

All that we have seen of the science of genetics in the preceding chapters has underlined one important point: genetics is a package deal. In other words, from the perspective of the individual organism, the genetic variation that enables the organism to exist in the first place is of exactly the same type as that which causes the organism to have some genetic disease, or to die young. In fact, the more you look at biological life, the more you realize that every item that we might count as a "plus" also has an inevitable "minus." The genetic variation that enables evolution to occur, entailing that every single human being is a unique individual, is also the variation that can result in cancer. Organisms can only survive if they have some way of understanding what is happening in their own bodies and in the environment around them. For sentient beings like ourselves, that means pain. The higher the level of conscious awareness, the greater the experience of pain. Without pain we would not survive for long. Those rare individuals born with genetic mutations that mean they feel no pain often die young, unless the problem is correctly diagnosed at a very early age.

We can build a great table, with two columns, in which all the "pluses" lined up on one side are finely balanced by all the "minuses" on the other. Eating food seems a pretty good plus, although it has a minus: it increases the dangerous form of oxygen ("free oxygen radicals") inside cells, so increasing the chance of DNA damage. All cells (that have a nucleus) are programmed to self-destruct, should that be necessary, by apoptosis. This self-destruction is essential and positive during the development of the brain, nerves, muscles, and many other organs. Conversely, when the process of apoptosis becomes dysfunctional, then it can lead to cancer.

Indeed, life itself is impossible without death. We are all part of great, long food chains, and the dead are constantly making way for the living. Imagine a world in which there was no death. The planet would soon be crammed with bacteria, not to speak of all the other organisms. Even if we rendered humans alone immortal in our thought experiment, the planet would only take a few thousand years to become as crowded as a New York subway during rush hour. The fact of the matter is that we are all on the great escalator of life, and death is an inevitable part of that process.

So why couldn't God just create a world in which all the things we don't like were simply eliminated, and all the nice things were maintained? The problem, of course, comes from some very practical, biological considerations. If it is a great good that conscious beings should come into being with minds that are able to make moral choices, especially to choose willingly and without any constraints to come into relationship with the personal creator God of the universe, then the only way we know about by which this could occur is by means of carbon-based life that renders the existence of such conscious, rational minds feasible by means of a long evolutionary history. Furthermore, it is only within a world of "nomic regularity" that such beings could exist, and indeed that any living thing can exist.

By the use of the term "nomic regularity," philosophers wish to draw attention to the lawlike behavior of the properties of matter.

If things didn't behave consistently, then you wouldn't know where you were, and nor would anything else. A world of consistent cause and effect means that if a big fish moves with a certain speed at a certain angle, then it will catch the little minnow for supper. It means that we can play baseball and do algebra. It means that if the sheriff shoots the villain, then the bullet goes in a straight line and doesn't suddenly move sideways around the criminal's head. In other words, God is not a magician. God has created a world of nomic regularity in which things behave in an orderly and predictable fashion. If it were not so, living things could not exist, evolution wouldn't happen, and we wouldn't be here.

It might be objected that this line of argument is constrained to the kind of life forms that we know about, that is, carbon-based life. It is possible that in other parts of the universe, life exists based on other chemical elements, such as silicon. We cannot rule this out, but it seems unlikely because we do know that the chemistry of the universe is remarkably similar every way we look (and radio telescopes enable us to look very far indeed). If the chemistry is the same, then it is very likely that the biochemistry is very similar also, and there are very good chemical reasons why life should be based around carbon. Disappointingly for sci-fi enthusiasts, it seems likely that life forms in other planets might be boringly similar to what we have here on planet Earth.

But the objection might still be pressed that the argument so far has been based very much on what we know. Surely God must know many things that we don't know, and therefore has a way of bringing about all the goods that characterize this planet's life forms, and yet without the kinds of downside that we have been considering. As it happens, this is exactly what Christians believe, since they maintain that this present existence is the preparation time for a future "new heavens and new earth" in which the particular downsides that we have been considering no longer exist.[10] Indeed, this future existence gives meaning to the present; if the present existence represented the end of the story, then it would indeed be difficult to mount a satisfactory response to the objection that suffering subverts the notion of a God of love. But if life in an eternity of bliss finally renders the present life as the mere blink of an eyelid by comparison, then that certainly does cast a different light on things.

But if that is the case, then why can't we just go there immediately without passing through this vale of suffering along the way? The Christian answer to that is that God only wills to have people in his "new heavens and new earth," a totally new mode of existence, yet with continuity with the present order of things, who willingly choose to be there. The character of love is such that it woos but never coerces. The only way to generate conscious, rational beings with the ability to freely choose to respond to God's love (or not) is, it seems, to generate carbon-based life that can, as a matter of fact, generate creatures with precisely the required properties by a long evolutionary process.

Now, of course, we cannot possibly know that this last claim is in fact true. There is no way in which we could know the mind of God in order to know all the possible ways there might be to bring about the existence of individuals able to make freely willed responses to God's love. So at this stage a simple parable might be helpful.[11] Let us imagine that my wife is going on a trip to Brighton (a town on the south coast of England) to stay for a few days with some friends. During this period I hear a report from a friend of mine that he happened to be in a restaurant in London where he saw my wife holding hands with a man whom he could not identify. On the face of it, this does not look good. On the other hand, we have been married for forty years, during which time my wife has demonstrated only love and faithfulness, and therefore it would be totally out of character for her to be unfaithful. Furthermore, I happen to know that she has an uncle who lives in London who has recently been seriously ill. Could it be, I wonder, that her uncle has suddenly taken a turn for the worse, and she therefore aborted the time with her friends in Brighton in order to take her dying uncle out for a last meal in London? I also know that she forgot her cell phone, because I saw it lying in the kitchen, and so she might not have been able to call to explain the change in plans.

Notice that I do not need to know the actual truth about the situation in order to have a reasonable justification that maintains my belief in my wife's faithfulness. It is only necessary that I can envisage a scenario that explains the reported observation.

This parable seems to provide at least an inkling of the relationship between the person convinced of God's love and the observed data of the natural evils that we have been discussing. We have no way of knowing the actual truth of the situation because that would entail knowing the mind of God, which we cannot possibly know. But Christians certainly know that God's love has been demonstrated to them in the most powerful way possible by the fact of God's incarnation in the world in the person of Christ, whose subsequent sacrifice on the cross as a penalty for our sin enables redeemed humanity to come to know God in a personal way by freely willed choice. Furthermore, Christians will also have experienced themselves the love of God conveyed to their own lives in all kinds of ways, perhaps for forty years or longer. Given this fact, they find themselves in the position of the puzzled husband in the parable. It might well be the case that the *only* way in which conscious beings can be created with genuine free will is via the pathway of costly carbon-based life with all its attendant "pluses" and "minuses," complete with all the genetic variation that represents such a two-edged sword. There is no way of knowing whether that is in fact the case. But the reflection that it just might be so is sufficient to keep the believer in the love of God going because, after all, the data that demonstrate God's love are so overwhelming that ambiguous data that may or may not count to the contrary can safely be put on the back burner until things become clearer. As it happens, scientists adopt this strategy all the time: a theory is believed because it provides a good explanation for a wide range of results, and anomalous data that don't fit the theory are generally put on the back burner until new techniques or approaches allow them to be reassessed at some time in the future.

The believer in the love of God therefore finds herself in a somewhat analogous position to the physicist convinced of the truth of Schrödinger's equation. This is the equation that describes and predicts the quantum behavior of matter. The equation has withstood every experimental test that has been thrown at it so far: it provides a brilliant insight into the properties of matter. But conceptually no one can quite get their head around the quantum behavior of matter because so much of it is counterintuitive. Perhaps one day there will be new insights that enable us to see how quantum mechanics can be conceptualized in more imaginable ways. But that day is not with us yet. Key pieces of the puzzle are still missing.

The believer in the love of God in like manner has pieces of the puzzle missing, but lives in faith that the pieces will become clearer one day. For the present, just because a few pieces are missing, there seems no good reason not to embrace the main outline of the scenery revealed by the assembled jigsaw pieces. A few pieces can be missing from a jigsaw puzzle, but you can still get the general picture pretty clearly.

In certain ways, also, contemporary genetics has made the challenge of suffering less subversive to the notion of a God of love, for we can now see that it is no arbitrary or useless aspect of biological existence, but an integral part of the whole process in which variant DNA lies at the heart of the existence of all living things. Perhaps it seems unfair that the costs of nomic regularity should be borne more by some rather than others, marked by a youthful cancer or debilitating genetic disease. These are certainly not easy challenges to address, either theologically or pastorally. But at least we can be sure (aside from induced DNA damage caused by lifestyle choices such as smoking) that the genetic disease has nothing to do with the sufferer's personal life or moral actions, but is simply part and parcel of being human. For we are all genetically programmed to die, some sooner than others, and that should help to focus all of our minds both on the value of our present lives, and on whether we are really prepared for the life that is to come.

Genetics—How Far Should We Go?

Just as reflection on genetics has helped us see that pain and suffering are not gratuitous, but are part and parcel of the larger scheme of things, so from a practical perspective it is also our new insights into genetics that have, as we saw in chapter 11, the huge potential to help and to heal. This raises the ethical question: how far should we go in using DNA technologies to not only heal but maybe also enhance human existence?

Curiously it was the late author Michael Crichton (he of *Jurassic Park* fame) who picked up something of my own perspective on the matter in his novel *Next*.[12] In the novel we have the "handsome and assured" Bellarmino who makes the following statement in a presentation at a conference titled "God's Plan for Mankind in Genetic Science": "My thanks to Congressman Waters, and to all of you for coming today. Some of you may wonder how a scientist—especially a genetic scientist—can reconcile his work with the word of God. But as Denis Alexander points out, the Bible reminds us that God, the Universal Creator, is separate from His creation but that He also actively sustains it moment to moment. Thus God is the creator of DNA, which underlies the biodiversity of our planet" (and so on, for another two pages).

Now it would be unfair to expect a novelist, while writing a work of fiction, to reproduce with complete accuracy the thinking of a geneticist in a far-off land. As should be clear from the discussion above, I do not believe that "God is the creator of DNA" in perhaps quite the way that Crichton's fictional character envisaged, but certainly DNA is an integral part of the "creation," that which exists by God's say-so. Furthermore, within the framework provided by Christian theism, we are given the mandate to care for DNA just as much as any other part of God's good creation: it is all part of being made "in God's image," with all the attendant responsibilities to care for God's earth that this entails. As mentioned at the outset of this chapter, we—uniquely among all organisms—care about other aspects of the created order.

In practice, as far as genetics is concerned, our primary responsibility is rather clear: to use our genetic knowledge in the healing of disease. As we have seen, there is room for some cautious optimism concerning the future of genetic therapies for individual sufferers. The hopes and expectations of the 1960s are finally being fulfilled more than half a century later. But what about the manipulation of the human germ line to eliminate all genetic diseases? Wouldn't that be much better than treating each individual patient? At present there is no risk-free technology that would allow the replacement of a defective gene in the human genome with the normal copy. Besides, tampering with the germ-line genome is banned in most countries, and its alteration would never get past medical ethical committees.

A further practical fact changes the nature of the discussion considerably, and that is the practice of preimplantation diagnosis. Preimplantation diagnosis involves in vitro fertilization (IVF) followed by growth of the embryo to the stage at which it contains four to eight cells. One or two cells can then be removed without damaging the embryo and defective genes identified. Embryos carrying the defective gene are discarded and only the healthy embryos are implanted in the mother. The technique bypasses the need to manipulate the genome, with all the attendant possible risks that this entails, and focuses instead on the selection of embryos without the mutant gene. In practice preimplantation diagnosis is normally offered to parents where there is a known history of debilitating genetic disease. The IVF procedure involved is in itself not risk-free and involves some considerable discomfort to the mother. Some believe that the embryo wastage that is involved in IVF is sufficient grounds for rejecting such a procedure, although it should be remembered that embryo wastage is an inevitable aspect of normal human procreation, in which up to 80 percent of fertilized eggs will never go to full term.

However efficient one might be at eliminating genetic diseases from the human gene pool using such approaches, the techniques would never be totally successful for the simple reason that new mutations are constantly occurring, so there would always be diseases that appear "out of the blue." Nevertheless, preimplantation diagnosis is clearly of huge benefit to those parents with a family history of genetic disease, and has some great advantages in comparison with the alternative scenario of manipulating the human germ line. Like all beneficial technologies, however, it is also open to abuse, and is in danger of being used to prevent the birth of babies with treatable medical conditions, or even with the "wrong" sex, or "wrong" eye or hair color. Preimplantation diagnosis was developed for the prevention of lethal diseases, especially those in which the affected child dies painfully within the first decade of life, and in those contexts the approach would appear to be ethically most appropriate.

What about enhancing the human germ line so that offspring display some qualities preferred by their parents? The problematic nature of this question should be clear from the rest of this book. Let us not forget that even a relatively "simple" human trait such as height is regulated by 180 or more variant genes. When it comes to traits such as musical ability or intelligence, then the genetic contributions, if any, are so complex that any idea of genetic manipulation to achieve them is simply unrealistic. The complexity of the genome remains its own best protection against the aspirations of the "enhancers."

Human autonomy also becomes an important element in the discussion. Who is going to decide what is beneficial to the human genome? In a thought experiment, let us imagine parents who wish their son to be a successful basketball player. They therefore pay for genetic manipulation of his genome during the course of an IVF procedure, involving dozens of different genes, to ensure that he will at least be of sufficient height to give him an advantage in becoming a successful player (in practice, this is a sci-fi scenario way beyond our present capabilities). The genetic manipulation, taken together with the parents' hopes and expectations, would seem to be an unwelcome intrusion into the life of an autonomous individual. And in any case, as already noted, being tall is advantageous but not essential for being a successful basketball player. It might just be the parents' luck if their offspring decided to be a geneticist. Human free will is perhaps an even greater protection from enhancement than the complexity of the genome itself.

So the answers to the question "How far should we go in manipulating DNA?" turn out to be rather prosaic. The scope for healing the individual sufferer is very great indeed, limited at present only by the inefficiency of our therapies. Yet the potential for changing the human germ line seems slight indeed, and not only slight but unnecessary.

Genetics and the Purpose of Life

I write this (far from Cambridge, U.K., on vacation, it has to be admitted) looking down over the luxuriant green valleys and wooded peaks of the Adirondack Mountains of upstate New York. My mind goes back to that time before 3.8 billion years ago when the planet was a lifeless collection of rocks and empty oceans, and I try with some difficulty to picture the same scene before me, yet without life of any kind. Somehow, in ways that we as yet only dimly understand, the abundance of life emerged from inanimate matter, and DNA has turned out to be the star player in life forms ever since, giving rise to all the luxuriant diversity that makes this planet such an amazing place in which to live.

We are so used to the fact that DNA and genetic diversity emerged from inanimate matter that we forget how very odd this really is. The very existence of the 3.8-billion-year evolutionary history, in which genetics has played a central part, seems to require some kind of explanation. Looking at a lifeless desert, the DNA double-helix is not the first entity that you might expect from such an apparently unpromising landscape. In the rest of this book we have been looking at proximal explanations for things, because that is what science is about. But this scientific level of explanation by no means excludes some overall purpose for the narrative taken as a whole. It is the existence of the narrative itself that stimulates our curiosity.

This book began with a personal biographical narrative, with some urine turning black as the Israeli tanks rolled up the road from Tyre and Sidon toward Beirut. As it happens, we were evacuated three times from Beirut during those troubled years, but were able to return after the first two occasions, the third being the coup de grace. I am sometimes asked why I switched research fields from human genetics to immunology and cancer. My answer is simple: "Because President Reagan decided to bomb Libya." In June 1986 planes took off from a U.S. base near Cambridge (U.K.) in a crazy mission to try and assassinate Colonel Ghaddafi from the air. The mission was a failure, and the Arab world was naturally incensed. Until that time the kidnappers of West Beirut had left the tiny British community that remained there reasonably untouched. This situation now changed because of British collusion with the failed mission; three Westerners were killed in retaliation for the raid, including two Brits. Within two days we were evacuated for our third and last time, experiments left half-finished on the laboratory bench, furniture hurriedly sold to friends who came to say good-bye, much-loved cats given away to good homes. Suddenly finding myself without a job and with a family to feed, I took the first research position that opened up back in the United Kingdom, which happened to be in the field of immunology and cancer. So President Reagan did indeed play a critical role in my future research career.

All of our lives represent an immensely complex interweaving of myriad events that together help construct our biographies. Without the language of the genes, we wouldn't be here: no history, no fraught politics, no science, no art, just a boring rocky planet with big oceans and no life. But if we told the genetic narrative as if it were the only one that matters, then certainly we would be missing a huge slice of what most people find to be the most important aspects of their lives: the friendships, the careers, the family life, the experience of beauty, the arts, love, and sex. Of course, we can always maintain that, in the final analysis, this is all a "tale told by an idiot" without any ultimate meaning or purpose. But for myself, I think not. If something looks and feels as if it has some purpose, then the most reasonable inference to adopt would seem to be that this is because it really does have a purpose; it seems perverse to conclude otherwise. The genetic data do not *force* that conclusion upon us: we can interpret them otherwise. The biographies woven by the decisions and events of our own lives do not *force* that conclusion upon us either; for some of us, life's contingencies may make us look more in other directions.

But the judicious observer cannot fail to be impressed by the fact of the existence of conscious, intelligent agents on planet Earth, made possible by the tight constraints of our genetic history, without whose existence there would be no awareness or appreciation of the remarkably coherent properties of this mathematically elegant and intelligible universe. The narrative provided by Christian theology renders these observations coherent: the emergence of conscious personhood is what you expect in a universe that is the creation of a personal God with intentions and purposes for the world. A tightly constrained genetic history is likewise consistent with God's purpose in bringing such beings into existence. Most remarkable of all, the biological constitution of such beings is of such a kind that they can freely respond, or not respond, to God's love, without the slightest coercion, and with full responsibility for their actions.

The language of the human genome specifies the building of organisms, each one unique, that can make genuinely free decisions for which they can be held accountable. Perhaps that single fact more than any other underlines the remarkable way in which genetics itself points us to a purpose and destiny that lies beyond genetics.

Notes

Preface

1. F. F. Stenn et al., "Biochemical Identification of Homogentisic Acid Pigment in an Ochronotic Egyptian Mummy," *Science* 197 (1977): 566–68.

Chapter 1

1. Benjamin A. Pierce, *Genetics: A Conceptual Approach,* 3rd ed. (New York: W. H. Freeman, 2008), 7–8.
2. *Corpus Hippocraticum* VII, 471–75. Also see Ute Deichmann, "Gemmules and Elements: On Darwin's and Mendel's Concepts and Methods in Heredity," *Journal of General Philosophy* 41 (2010): 85–112.
3. Cited in Peter Vorzimmer, "Inheritance through Pangenesis," *Dictionary of the History of Ideas* (The Electric Text Center at the University of Virginia Library, 2003).
4. F. Rosner, "Hemophilia in Classic Rabbinic Texts," *Journal of the History of Medicine and Allied Sciences* 49 (1994): 240–50.
5. Vítězslav Orel, *Gregor Mendel: The First Geneticist* (New York: Oxford University Press, 1996).
6. Brian J. Ford, "Brownian Movement in Clarkia Pollen: A Reprise of the First Observations," *Microscope* 40 (1992): 235–41. Robert Brown is most famous for discovering what is now known as "Brownian motion" in physics.
7. Charles Darwin, *Variation of Plants and Animals under Domestication,* vol. 2 (London: John Murray, 1868), 350.
8. Edward J. Larson, *Evolution* (New York: Modern Library, 2006).
9. Ibid., 163.
10. William Bateson, *Mendel's Principles of Heredity* (Cambridge: Cambridge University Press, 1902).
11. Archibald E. Garrod, "The Incidence of Alkaptonuria: A Study in Chemical Individuality," *Lancet* 160 (1902): 1616–20. This classic article was reproduced in the *Yale Journal of Biology and Medicine* 75 (2002): 221–31.
12. In a personal letter to Adam Sedgwick, dated April 18, 1905.
13. Rene J. Dubos, *The Professor, the Institute, and DNA* (New York: Rockefeller University Press, 1976), 20–23.
14. James D. Watson and Francis H. Crick, "Molecular Structure of Nucleic Acids—A Structure for Deoxyribose Nucleic Acid," *Nature* 171 (1953): 737–38.

15. http://www.ornl.gov/sci/techresources/Human_Genome/medicine/assist.shtml. Accessed November 17, 2010.
16. No analogy is perfect. A spiral staircase suggests that the base-pairs circle around the central axis, whereas in DNA the central axis passes through the base-pairs. Apart from that, the staircase picture works quite well.
17. http://vega.sanger.ac.uk/Homo_sapiens/mapview?chr=1. Accessed November 17, 2010.
18. Two further rare amino acids known as selenocysteine and pyrrolysine that are used only in special contexts.
19. The code is given at http://users.rcn.com/jkimball.ma.ultranet/BiologyPages/C/Codons.html. Accessed August 22, 2010.

Chapter 2

1. Francis H. Crick, "On Protein Synthesis," *Society of Experimental Biology* 12 (1958): 139–63.
2. http://www.ncbi.nlm.nih.gov/Web/Newsltr/FallWinter02/longest.html. Accessed August 14, 2009.
3. Benjamin Lewin, *Genes IX* (Sudbury, MA: Jones and Bartlett, 2008), 137.
4. Ibid., 135.
5. http://rice.plantbiology.msu.edu/pseudomolecules/info.shtml. Accessed August 18, 2009.
6. A group of twenty-five biologists using computer programs to identify genes spent two days trying to come up with a definition of a gene with which they all agreed. Obviously if you do not know what a gene is, finding one is difficult! As the leader of the consortium, Karen Eilbeck from the University of California at Berkeley reported, "We had several meetings that went on for hours and everyone screamed at each other," before they came up with a definition that could accommodate everyone's demands. Their definition of a gene was as follows: "A locatable region of genomic sequence, corresponding to a unit of inheritance, which is associated with regulatory regions, transcribed regions, and/or other functional sequence regions." H. Pearson, "Genetics: What Is a Gene?" *Nature* 441 (2006): 398–401. My definition says basically the same thing in less flowery language.
7. C. Pitnick, G. S. Spicer, and T. A. Markow, "How Long Is a Giant Sperm?" *Nature* 375 (1995): 109.
8. L. C. Ryner et al., "Control of Male Sexual Behavior and Sexual Orientation in Drosophila by the *Fruitless* Gene," *Cell* 87 (1996): 1079–89.
9. D. Schmucker et al., "*Drosophila* Dscam Is an Axon Guidance Receptor Exhibiting Extraordinary Molecular Diversity," *Cell* 101 (2000): 671–84.
10. Y. Barash et al., "Deciphering the Splicing Code," *Nature* 465 (2010): 53–59; H. Ledford, "The Code within the Code," *Nature* 465 (2010): 16–17.
11. S. Griffiths-Jones, "Annotating Noncoding RNA Genes," *Annual Review of Human Genomics and Human Genetics* 8 (2007): 279–98; D. H. Chitwood and M. C. P. Timmermans, "Small RNAs Are on the Move," *Nature* 467 (2010): 415.
12. C. P. Ponting and G. Lunter, "Human Brain Gene Wins Genome Race," *Nature*

443 (2006): 149–50; K. S. Pollard et al., "An RNA Gene Expressed during Cortical Development Evolved Rapidly in Humans," *Nature* 443 (2006): 167–72.

13. http://www.ensembl.org/Homo_sapiens/Info/StatsTable. Accessed November 18, 2010.
14. H. Van Bakel et al., "Most 'Dark Matter' Transcripts Are Associated with Known Genes," *PLOS Biology* 8 (2010): e1000371.
15. J. E. Wilusz, H. Sunwoo, and D. L. Spector, "Long Noncoding RNAs: Functional Surprises from the RNA World," *Genes and Development* 23 (2009): 1494–504; U. A. Ørom et al., "Long Noncoding RNAs with Enhancer-like Function in Human Cells," *Cell* 143 (2010): 46.
16. American friends gave me this analogy. I have no idea what it means in sport, but hopefully it conveys something of the science to the initiated. If I put an example from cricket in here, then I'd lose another segment of my readership. You can't win on these things.

Chapter 3

1. R. J. Sims and D. Reinberg, "Escaping Fates with Open States," *Nature* 460 (2009): 802–3; A. Gaspar-Maia et al., "Chd1 Regulates Open Chromatin and Pluripotency of Embryonic Stem Cells," *Nature* 460 (2009): 863–68.
2. Chd1 stands for Chromodomain helicase DNA binding protein 1.
3. Denis Noble, *The Music of Life* (New York: Oxford University Press, 2006).
4. A very accessible and more detailed account of such genes is given in Sean B. Carroll, *Endless Forms Most Beautiful* (London: Weidenfeld & Nicolson, 2005). See also R. K. Maeda and F. Karch, "The ABC of the BX-C: The Bithorax Complex Explained," *Development* 133 (2006): 1413–22.
5. M. Ronshaugen et al., "The Drosophila MicroRNA iab-4 Causes a Dominant Homeotic Transformation of Halteres to Wings," *Genes and Development* 19 (2005): 2947–52.
6. A. Stark et al., "A Single Hox Locus in *Drosophila* Produces Functional MicroRNAs from Opposite DNA Strands," *Genes and Development* 22 (2008): 8–13.
7. N. E. Stork, "World of Insects," *Nature* 448 (2007): 657–58.
8. G. E. Robinson, "Sociogenomics Takes Flight," *Science* 297 (2002): 204–5; G. E. Robinson, C. M. Grozinger, and C. W. Whitfield, "Sociogenomics: Social Life in Molecular Terms," *Nature Reviews Genetics* 6 (2005): 257–70.
9. C. R. Smith et al., "Genetic and Genomic Analyses of the Division of Labour in Insect Societies," *Nature Genetics* 9 (2008): 735–48.
10. L. Wilfert, J. Gadau, and P. Schmid-Hempel, "Variation in Genomic Recombination Rates Among Animal Taxa and the Case of Social Insects," *Heredity* 98 (2007): 189–97.
11. Sadly, though, not total protection. Colony Collapse Disorder in honeybee hives has recently been a major problem in various parts of the United States, leading to loss of 50 to 90 percent of all colonies. Colonies undergoing collapse are associated with a higher load of quite a range of pathogens. See http://www.sciencemag.org/cgi/content/abstract/1146498 and http://www.laboratoryequipment.com/News-honeybee-infections-mystery-081409.aspx. Accessed August 25, 2009.

12. E. Szathmáry and S. Számadó, "Language: A Social History of Words," *Nature* 456 (2008): 40–41.
13. T. Giraud, J. S. Pedersen, and L. Keller, "Evolution of Supercolonies: The Argentine Ants of Southern Europe," *Proceedings of the National Academy of Sciences USA* 99 (2002): 6075–79.
14. C. S. Lai et al., "A Forkhead-Domain Gene Is Mutated in a Severe Speech and Language Disorder," *Nature* 413 (2001): 519–23.
15. M. H. Dominguez and P. Rakic, "The Importance of Being Human," *Nature* 462 (2009): 169–70.
16. S. Haesler et al., "Incomplete and Inaccurate Vocal Imitation after Knockdown of FoxP2 in Songbird Basal Ganglia Nucleus Area X," *PLoS Biology* 5 (2007): e321; doi:10.1371/journal.pbio.0050321.
17. L. Gang et al., "Accelerated FoxP2 Evolution in Echolocating Bats," *PLoS ONE* 2 (2007): e900; doi:10.1371/journal.pone.0000900.
18. J. Krause et al., "The Derived FOXP2 Variant of Modern Humans Was Shared with Neandertals," *Current Biology* 17 (2007): 1908–12; R. E. Green et al., "A Draft Sequence of the Neandertal Genome," *Science* 328 (2010): 710–22.
19. E. Fujita et al., "Ultrasonic Vocalization Impairment of Foxp2 (R552H) Knockin Mice Related to Speech-language Disorder and Abnormality of Purkinje Cells," *Proceedings of the National Academy of Sciences USA* 105 (2008): 3117–22.
20. S. C. Vernes et al., "High-Throughput Analysis of Promoter Occupancy Reveals Direct Neural Targets of *FOXP2,* a Gene Mutated in Speech and Language Disorders," *American Journal of Human Genetics* 81 (2007): 1232–50.

Chapter 4

1. The answer is $10 \times 2^{72} = 4.7 \times 10^{22}$, or 47,000,000,000,000,000,000,000 bugs.
2. Benjamin Lewin, *Genes IX* (Sudbury, MA: Jones and Bartlett, 2008), 19.
3. E. D. Pleasance et al., "A Small-Cell Lung Cancer Genome with Complex Signatures of Tobacco Exposure," *Nature* 463 (2010): 184–90. In another study of a lung tumor from a male who reported smoking twenty-five cigarettes a day for fifteen years, more than 50,000 mutations in the cancer cells were reported, of which 392 were in protein-coding regions; see W. Lee et al., "The Mutation Spectrum Revealed by Paired Genome Sequences from a Lung Cancer Patient," *Nature* 465 (2010): 473–77.
4. E. D. Pleasance et al., "A Comprehensive Catalogue of Somatic Mutations from a Human Cancer Genome," *Nature* 463 (2010): 191–96.
5. The 1000 Genomes Project Consortium, "A Map of Human Genome Variation from Population-Scale Sequencing," *Nature* 467 (2010): 1061–73; J. C. Roach, et al., "Analysis of Genetic Inheritance in a Family Quartet by Whole-Genome Sequencing," *Science* 328 (2010): 636–39.
6. Pete Moore, *Enhancing Me: The Hope and the Hype of Human Enhancement* (Hoboken, NJ: Wiley, 2008), 194–98. The hormone EPO is produced in the kidneys and regulates red blood cell production in the bone marrow. Red blood cells carry oxygen around the body, thereby giving athletic advantage. But having too many red blood cells is medically dangerous. In the four years after EPO

became available in Europe, twenty cyclists died of sudden and unexpected cardiac problems.

7. R. Cordaux and M. A. Batzer, "The Impact of Retrotransposons on Human Evolution," *Nature Reviews Genetics* 10 (2009): 691–703.
8. R. Cordaux et al., "Estimating the Retrotransposition Rate of Human *Alu* Elements," *Gene* 373 (2006): 134–37. This figure is based on the frequency of new *Alu* and LINE-1 inserts, each at one in twenty births.
9. J. M. Chen et al., "A Systematic Study of LINE–1 Endonuclease-dependent Retrotranspositional Events Causing Human Genetic Disease," *Human Genetics* 117 (2005): 411–27; P. A. Callinan and M. A. Batzer, "Retrotransposable Elements and Human Disease," in J-N Volff, *Genome and Disease, Genome Dynamics,* Basel, Switzerland: Karger, 1 (2006): 104–15; M. Mine et al., "A Large Genomic Deletion in the *PDHX* Gene Caused by the Retrotranspositional Insertion of a Full-length LINE–1 Element," *Human Mutation* 28 (2007): 137–42.
10. N. G. Coufal, "L1 Retrotransposition in Human Neural Progenitor Cells," *Nature* 460 (2009): 1127–31.
11. A. Fortna et al., "Lineage-Specific Gene Duplication and Loss in Human and Great Ape Evolution," *PLoS Biology* 2 (July 2004): e207; doi:10.1371/journal.pbio.0020207.
12. G. H. Perry et al., "Diet and the Evolution of Human Amylase Gene Copy Number Variation," *Nature Genetics* 39 (2007): 1256–60.
13. J. Randerson, "Record Breaker," *New Scientist*, June 8, 2002.
14. D. Giardino et al., "De Novo Balanced Chromosome Rearrangements in Prenatal Diagnosis," *Prenatal Diagnosis* 29 (2009): 257–65. "Balanced" variation occurs when a stretch of one chromosome is exchanged with a stretch of equivalent length from another chromosome.
15. Database of Genetic Variants, http://projects.tcag.ca/variation. Accessed March 3, 2010.
16. S. D. Turner and D. R. Alexander, "Fusion Tyrosine Kinase Mediated Signalling Pathways in the Transformation of Haematopoietic Cells," *Leukemia* 20 (2006): 572–82. In such cases the new gene formed is known as an "oncogene."
17. R. G. Walters et al., "A New Highly Penetrant Form of Obesity Due to Deletions on Chromosome 16p11.2," *Nature* 463 (2010): 671–75.
18. F. Zhang et al., "Complex Human Chromosomal and Genomic Rearrangements," *Trends in Genetics* 25 (2009): 298–307.
19. G. Hamilton, "The Gene Weavers," *Nature* 441 (2006): 683–85.
20. N. De Parseval et al., "Survey of Human Genes of Retroviral Origin: Identification and Transcriptome of the Genes with Coding Capacity for Complete Envelope Proteins," *Journal of Virology* 77 (2003): 10414–22.
21. Ibid.
22. S. Mi et al., "Syncytin Is a Captive Retroviral Envelope Protein Involved in Human Placental Morphogenesis," *Nature* 403 (2000): 785–89; S. Blaise et al., "Genome Wide Screening for Fusogenic Human Endogenous Retrovirus Envelopes Identifies Syncytin 2, a Gene Conserved on Primate Evolution," *Proceedings of the National Academy of Sciences USA* 100 (2003): 13013–18. Note that human syncytin-1 and -2 are different from murine syncytin A and

B; the human and murine viruses represent independent insertions. This nice example of the evolutionary convergence is discussed later.

23. M. Caceres and J. W. Thomas, "The Gene of Retroviral Origin Syncytin 1 Is Specific to Hominoids and Is Inactive in Old World Monkeys," *Journal of Heredity* 97 (2006): 100–106.
24. For a further interesting example, see the report on the incorporation of the Bornavirus into many mammalian genomes, occasionally to produce functional genes: C. Feschotte, "Bornavirus Enters the Genome," *Nature* 463 (2010): 39–40.
25. P. J. Keeling and J. D. Palmer, "Horizontal Gene Transfer in Eukaryotic Evolution," *Nature Reviews Genetics* 9 (2008): 605–18.
26. T. Dagan et al., "Modular Networks and Cumulative Impact of Lateral Transfer in Prokaryote Genome Evolution," *Proceedings of the National Academy of Sciences USA* 105 (2008): 10039–44.
27. N. D. Kristof, "The Spread of Superbugs," *New York Times*, March 6, 2010.
28. N. Goldenfeld and C. Woese, "Biology's Next Revolution," *Nature* 445 (2007): 369.
29. J. Qin et al., "A Human Gut Microbial Gene Catalogue Established by Metagenomic Sequencing," *Nature* 464 (2010): 59–65.
30. J. O. Andersson, "Lateral Gene Transfer in Eukaryotes," *Cellular and Molecular Life Sciences* 62 (2005): 1182–97.
31. A. M. Nedelcu, et al., "Adaptive Eukaryote-to-Eukaryote Lateral Gene Transfer: Stress-related Genes of Algal Origin in the Closest Unicellular Relatives of Animals," *Journal of Evolutionary Biology* 21 (2008): 1852–60.
32. First identified in mosquitoes in 1924 by M. Hertig and S. B. Wolbach.
33. J. C. D. Hotopp et al., "Widespread Lateral Gene Transfer from Intracellular Bacteria to Multicellular Eukaryotes," *Science* 317 (2007): 1753–56.

Chapter 5

1. Francisco J. Ayala, *Darwin's Gift to Science and Religion* (Washington, DC: Joseph Henry Press, 2007); Jerry A. Coyne, *Why Evolution Is True* (New York: Oxford University Press, 2009); Richard Dawkins, *The Greatest Show on Earth: The Evidence for Evolution* (New York: Bantam Press, 2009).
2. A more detailed explanation of the various nuances of the term "fitness" in evolutionary biology may be found in H. A. Orr, "Fitness and Its Role in Evolutionary Genetics," *Nature Reviews Genetics* 10 (2009): 531–39.
3. Stephen Jay Gould archive, http://www.stephenjaygould.org/people/john_haldane.html. Accessed November 19, 2010.
4. John B. S. Haldane, *The Causes of Evolution* (London: Longmans, Green, 1932), 32.
5. See note 2, this chapter.
6. Richard Dawkins, *The Selfish Gene* (Oxford: Oxford University Press, 1976).
7. L. D. Hurst, "Genetics and the Understanding of Selection," *Nature Reviews Genetics* 10 (2009): 83–93.
8. J. E. Barrick et al., "Genome Evolution and Adaptation in a Long-Term Experiment with *Escherichia Coli*," *Nature* 461 (2009): 1243–49; T. Chouard, "Revenge of the Hopeful Monster," *Nature* 463 (2010): 864–67.

9. M. Pagel et al., "Large Punctuational Contribution of Speciation to Evolutionary Divergence at the Molecular Level," *Science* 314 (2006): 119–21.
10. S. Paterson et al., "Antagonistic Coevolution Accelerates Molecular Evolution," *Nature* 464 (2010): 275–78.

Chapter 6

1. For example, see F. D. Ciccarelli, "Toward Automatic Reconstruction of a Highly Resolved Tree of Life," *Science* 311 (2006): 1283–87. Note the word "Toward" in the title. This is work in progress, and such "trees" or "bushes" (my preferred word) will become increasingly accurate as time goes on and more genomes are sequenced. The "tree" in this particular paper was based on 31 universal protein families and covered the genome sequences from 191 different species.
2. S. Carroll, "Chance and Necessity: The Evolution of Morphological Complexity and Diversity," *Nature* 409 (2001): 1102–9.
3. J. L. Boore, L. L. Daehler, and W. M. Brown, "Complete Sequence, Gene Arrangement, and Genetic Code of Mitochondrial DNA of the Cephalochordate Branchiostoma floridae (Amphioxus)," *Molecular Biology and Evolution* 16 (1999): 410–18.
4. I. Miranda, R. Silva, and M. A. S. Santos, "Evolution of the Genetic Code in Yeasts," *Yeast* 23 (2006): 203–13.
5. I. Miranda et al., "A Genetic Code Alteration Is a Phenotype Diversity Generator in the Human Pathogen *Candida albicans*," *PLoS One* 2 (2007): e996.
6. C. Dennis, "Coral Reveals Ancient Origins of Human Genes," *Nature* 426 (2003): 744.
7. M. Kellis et al., "Proof and Evolutionary Analysis of Ancient Genome Duplication in the Yeast *Saccharomyces cerevisiae*," *Nature* 428 (2004): 617–24.
8. P. Dehal and J. L. Boore, "Two Rounds of Whole Genome Duplication in the Ancestral Vertebrate," *PLoS Biology* 3 (2005): e314.
9. J. M. Thomson et al., "Resurrecting Ancestral Alcohol Dehydrogenases from Yeast," *Nature Genetic* 37 (2005): 630–63.
10. Standing for *fibroblast growth factor receptor 1.*
11. Francisco J. Ayala and M. Coluzzi, "Chromosome Speciation: Humans, *Drosophila*, and Mosquitoes," *Proceedings of the National Academy of Sciences USA*, 102 (2005): 6535–42. For a review on speciation in plants, see M. J. Hegarty and S. J. Hiscock, "Hybrid Speciation in Plants: New Insights from Molecular Studies," *New Phytologist* 165 (2005): 411–23.
12. M. Ridley, *Evolution*, 3rd ed. (Oxford: Blackwell, 2004), 53.
13. C. A. Wu et al., "*Mimulus* Is an Emerging Model System for the Integration of Ecological and Genomic Studies," *Heredity* 100 (2008): 220–30.
14. M. A. F. Noor and J. L. Feder, "Speciation Genetics: Evolving Approaches," *Nature Reviews Genetics* 7 (2006): 851–61.
15. L. Spinney, "Dreampond Revisited," *Nature* 466 (2010): 174–75.
16. E. Verheyan et al., "Origin of the Superflock of Cichlid Fishes from Lake Victoria, East Africa," *Science* 300 (2003): 325–29.
17. M. Sanetra, "A Microsatellite-Based Genetic Linkage Map of the Cichlid Fish, *Astatotilapia burtoni* (Teleostei): A Comparison of Genomic Architec-

tures among Rapidly Speciating Cichlids," *Genetics* 182 (2009): 387–97; C. J. Allender et al., "Divergent Selection during Speciation of Lake Malawi Cichlid Fishes Inferred from Parallel Radiations in Nuptial Coloration," *Proceedings of the National Academy of Sciences USA* 100 (2003): 14074–79.

18. P. L. Oliver et al., "Accelerated Evolution of the *Prdm9* Speciation Gene across Diverse Metazoan Taxa," *PLoS Genetics* 5 (2009): doi:10.1371/journal.pgen.1000753
19. Niles Eldredge and Stephen J. Gould, "Punctuated Equilibria: An Alternative to Phyletic Gradualism," in *Models in Paleobiology*, ed. T. J. M. Schopf (San Francisco: Freeman Cooper, 1972), 82–115.
20. D. L. Stern and V. Orgogozo, "Is Genetic Evolution Predictable?" *Science*, February 6, 2009, 746–51.
21. G. A. Wray, "The Evolutionary Significance of Cis-regulatory Mutations," *Nature Reviews Genetics* 8 (2007): 206–16.
22. Y. F. Chan et al., "Adaptive Evolution of Pelvic Reduction in Sticklebacks by Recurrent Deletion of a *Pitx1* Enhancer," *Science* 327 (2010): 302–5.
23. N. Shubin et al., "Deep Homology and the Origins of Evolutionary Novelty," *Nature* 457 (2009): 818–23.
24. David J. Bottjer, "The Early Evolution of Animals," *Scientific American*, August 2005, 30–35.
25. B. Guo et al., "*Hox* Genes of the Japanese Eel *Anguilla japonica* and *Hox* Cluster Evolution in Teleosts," *Journal of Experimental Zoology, Part B, Molecular and Developmental Evolution* 314 (2010): 135–47.
26. N. Di Poi et al., "Changes in Hox Genes' Structure and Function during the Evolution of the Squamate Body Plan," *Nature* 464 (2010): 99–103.
27. K. Kamm et al., "Axial Patterning and Diversification in the Cnidaria Predate the Hox System," *Current Biology* 16 (2006): 920–26.
28. Simon Conway Morris, *Life's Solution: Inevitable Humans in a Lonely Universe* (Cambridge: Cambridge University Press, 2003).
29. C. J. Jeffery, "Moonlighting Proteins," *Trends in Biochemical Sciences* 24 (1999): 8–11; A. Aharoni et al., "The Evolvability of Promiscuous Protein Functions," *Nature Genetics* 37 (2005): 73–76.
30. G. Jones, "Molecular Evolution: Gene Convergence in Echolocating Mammals," *Current Biology* 20 (2009): R62–64.
31. D. M. Weinreich et al., "Darwinian Evolution Can Follow Only Very Few Mutational Paths to Fitter Proteins," *Science* 312 (2006): 111–14.
32. I. S. Povolotskaya and F. A. Kondrashov, "Sequence Space and the Ongoing Expansion of the Protein Universe," *Nature* 465 (2010): 922–26.

Chapter 7

1. R. A. Foley and M. Mirazon Lahr, "The Base Nature of Neanderthals," *Heredity* 98 (2007): 187–88.
2. R. E. Green et al., "A Draft Sequence of the Neanderthal Genome," *Science* 328 (2010): 710–22.
3. J. Krause et al., "The Complete Mitochondrial DNA Genome of an Unknown Hominin from Southern Siberia," *Nature* 464 (2010): doi:10.1038/nat

ure08976; R. Dalton, "Fossil Finger Points to New Human Species," *Nature* 464 (2010): 472–73; D. Reich et al., "Genetic History of an Archaic Hominin Group from Denisova Cave in Siberia," *Nature* 468 (2010): 1053–60.

4. I. McDougall et al., "Stratigraphic Placement and Age of Modern Humans from Kibish, Ethiopia," *Nature* 433 (2005): 733–36.
5. T. D. White et al., "Pleistocene *Homo sapiens* from Middle Awash, Ethiopia," *Nature* 423 (2003): 742–47; J. D. Clark et al., "Stratigraphic, Chronological and Behavioural Contexts of Pleistocene *Homo sapiens* from Middle Awash, Ethiopia," *Nature* 423 (2003): 747–52.
6. A useful account of the spread of humanity out of Africa can be found in D. Jones, "Going Global," *New Scientist* 27 (October 2007): 36–41.
7. B. Pakendorf and M. Stoneking, "Mitochondrial DNA and Human Evolution," *Annual Reviews of Genomics and Human Genetics* 6 (2005): 165–83.
8. The mitochondrial data are consistent with an African origin, but by themselves are not conclusive. See N. A. Rosenberg and M. Nordborg, "Genealogical Trees, Coalescent Theory, and the Analysis of Genetic Polymorphisms," *Nature Reviews Genetics* 3 (2002): 380–90. However, further strong support for the "Out of Africa" model of human origins has come from detailed study of human skull dimensions. See A. Manica et al., "The Effect of Ancient Population Bottlenecks on Human Phenotypic Variation," *Nature* 448 (2007): 346–48. For those interested in the way that computational methods are used to construct lineages based on genetic variation in populations, see P. Marjoram and S. Tavare, "Modern Computational Approaches for Analyzing Molecular Genetic Variation Data," *Nature Reviews Genetics* 7 (2006): 759–70.
9. M. A. Jobling and C. Tyler-Smith, "The Human Y Chromosome: An Evolutionary Marker Comes of Age," *Nature Reviews Genetics* 4 (2003): 598–612. Some of the uncertainty in dating comes from doubts about how much polygamy there has been in human reproduction over the millennia.
10. The Chimpanzee Sequencing and Analysis Consortium, "Initial Sequence of the Chimpanzee Genome and Comparison with the Human Genome," *Nature* 437 (2005): 69–87.
11. J. F. Hughes et al., "Chimpanzee and Human Y Chromosomes Are Remarkably Divergent in Structure and Gene Content," *Nature* 463 (2010): 536–44; L. Buchen, "The Fickle Y Chromosome," *Nature* 463 (2010): 149.
12. Cited in Buchen, "Fickle Y Chromosome," 149
13. P. Khaitovich et al., "Parallel Patterns of Evolution in the Genomes and Transcriptomes of Humans and Chimpanzees," *Science* 309 (2005): 1850–54.
14. S. J. Sholtis and J. P. Noonan, "Gene Regulation and the Origins of Human Biological Uniqueness," *Trends in Genetics* 25 (2009): 82–90.
15. http://www.ensembl.org/Homo_sapiens/Info/StatsTable. Accessed April 4, 2010.
16. Y. Ohta and M. Nishikimi, "Random Nucleotide Substitutions in Primate Non-Functional Gene for L-gulono-γ-lactone Oxidase, the Missing Enzyme in L-ascorbic Acid Biosynthesis," *Biochimica et Biophysica Acta* 1472 (1999): 408.
17. B. J. Vincent et al., "Following the LINEs: An Analysis of Primate Genomic

Variation at Human-Specific LINE-1 Insertion Sites," *Molecular Biology and Evolution* 20 (2003): 1338.

18. Chimpanzee Sequencing and Analysis Consortium, "Initial Sequence of the Chimpanzee Genome and Comparison with the Human Genome," *Nature* 437 (2005): 69–87. Note that these numbers exclude contributions from microsatellite DNA.
19. R. Gibbons et al., "Distinguishing Humans from Great Apes with AuYb8 Repeats," *Journal of Molecular Biology* 339 (2004): 721–29.
20. Useful reviews describing the use of genetic fossils in constructing and confirming evolutionary lineages may be found in G. Finlay, "*Homo Divinus*: The Ape that Bears God's Image," *Science and Christian Belief* 15 (2003): 17–40; G. Finlay, "Evolution as Created History," *Science and Christian Belief* 20 (2008): 67–90. The examples shown are taken from these reviews, but hundreds of examples could be provided if space allowed.
21. D. J. Hedges et al., "Differential *Alu* Mobilization and Polymorphism among the Human and Chimpanzee Lineages," *Genome Research* 14 (2004): 1068–75.
22. The example is from Finlay, "*Homo Divinus*."
23. R. Cordaux and M. A. Batzer, "The Impact of Retrotransposons on Human Evolution," *Nature Reviews Genetics* 10 (2009): 691–703.

Chapter 8

1. Francis Collins, "Has the Revolution Arrived?" *Nature* 464 (2010): 674–75; J. C. Venter, "Multiple Personal Genomes Await," *Nature* 464 (2010): 676–77.
2. S. Levy, "The Diploid Genome Sequence of an Individual Human," *PLOS Biology* 5 (2007): e254.
3. E. S. Lander et al., "Initial Sequencing and Analysis of the Human Genome," *Nature* 409 (2001): 860–921; International Human Genome Sequencing Consortium, "Finishing the Euchromatic Sequence of the Human Genome," *Nature* 431 (2004): 931–45.
4. http://www.ensembl.org/Homo_sapiens/Info/StatsTable. Accessed April 6, 2010.
5. D. F. Conrad et al., "Origins and Functional Impact of Copy Number Variation in the Human Genome," *Nature* 464 (2010): 704–12.
6. N. J. Fagundes et al., "Statistical Evaluation of Alternative Models of Human Evolution," *Proceedings of the National Academy of Sciences USA* 104 (2007): 17614–19.
7. B. Linz et al., "An African Origin for the Intimate Association between Humans and *Helicobacter pylori*," *Nature* 445 (2007): 915–18.
8. D. R. Bentley et al., "Accurate Whole Human Genome Sequencing Using Reversible Terminator Chemistry," *Nature* 456 (2008): 53–59.
9. R. Li et al., "Building the Sequence Map of the Human Pan-Genome," *Nature Biotechnology* 28 (2010): 57–63.
10. The 1000 Genomes Project Consortium, "A Map of Human Genome Variation from Population-Scale Sequencing," *Nature* 467 (2010): 1061–73.
11. S. C. Schuster, "Complete Khoisan and Bantu Genomes from Southern Africa," *Nature* 463 (2010): 943–47.

12. The uses and abuses of biology for ideological reasons from 1600 to the present day, including a history of the eugenics movement, can be found in Denis R. Alexander and Ronald L. Numbers, *Biology and Ideology: From Descartes to Dawkins* (Chicago: University of Chicago Press, 2010).
13. G. Barbujani and V. Colonna, "Human Genome Diversity: Frequently Asked Questions," *Trends in Genetics* 26 (2010): 285–95.
14. Ibid.
15. www.hapmap.org; "The International HapMap Project," *Nature* 426 (2003): 789–96. The "1000 Genomes Project" was also launched in early 2008—the plan to sequence the complete genomes of one thousand people. See The 1000 Genomes Project Consortium, "A Map of Human Genome Variation from Population-Scale Sequencing," *Nature* 467 (2010): 1061–73.
16. E. C. Hayden, "Similar, Yet So Different," *Nature* 449 (2007): 762–63.
17. J. K. Pickrell et al., "Signals of Recent Positive Selection in a Worldwide Sample of Recent Populations," *Genome Research* 19 (2009): 826–37.
18. B. T. Lahn and L. Ebenstein, "Let's Celebrate Human Genetic Diversity," *Nature* 461 (2009): 726–28.
19. R. Dalton, "Icy Resolve," *Nature* 463 (2010): 724.
20. D. M. Lambert and L. Huynen, "Face of the Past Reconstructed," *Nature* 463 (2010): 739–40; M. Rasmussen et al., "Ancient Human Genome Sequence of an Extinct Palaeo-Eskimo," *Nature* 463 (2010): 757–62.
21. D. Reich et al., "Reconstructing Indian Population History," *Nature* 461 (2009): 489–95.

Chapter 9

1. H. Pearson, "Human Genetics: One Gene, Twenty Years," *Nature* 460 (2009): 164–69.
2. http://www.ncbi.nlm.nih.gov/omim. This database started life as a series of twelve print volumes produced by Dr. Victor McCusick, initiated in the early 1960s. The online version is now freely available and much easier to keep updated.
3. As of April 19, 2010.
4. http://www.hgmd.cf.ac.uk/ac/index.php maintained by the Institute of Medical Genetics, Cardiff, U.K. Accessed May 17, 2010.
5. P. M. Quinton, "Chloride Impermeability in Cystic Fibrosis," *Nature* 301 (1983): 421–22.
6. Francis Collins, *The Language of Life: DNA and the Revolution in Personalized Medicine* (New York: HarperCollins, 2010).
7. Cited in H. Pearson, "Human Genetics: One Gene, Twenty Years," *Nature* 460 (2009): 164–69.
8. In a paper written when George Huntington was only twenty-two years old, one year after graduating from Columbia University Medical School. In fact, Huntington recognized dominant inheritance nearly three decades before Mendelian inheritance was rediscovered at the turn of the century, writing that when "either or both the parents have shown manifestations of the disease . . . , one or more of the offspring almost invariably suffer from the disease. . . . But

if by any chance these children go through life without it, the thread is broken and the grandchildren and great-grandchildren of the original shakers may rest assured that they are free from the disease."

9. A. R. La Spada and J. P. Taylor, "Repeat Expansion Disease: Progress and Puzzles in Disease Pathogenesis," *Nature Reviews Genetics* 11 (2010): 247–58.
10. V. Dion and J. H. Wilson, "Instability and Chromatin Structure of Expanded Trinucleotide Repeats," *Trends in Genetics* 25 (2009): 288–97.
11. A. Aguzzi and T. O'Connor, "Protein Aggregation Diseases: Pathogenicity and Therapeutic Perspectives," *Nature Reviews Drug Discovery* 9 (2010): 237–48.
12. DGCR8, if you really want to know.
13. K. R. Chi, "Hit or Miss?" *Nature* 461 (2009): 712–14.
14. H. L. Allen et al., "Hundreds of Variants Clustered in Genomic Loci and Biological Pathways Affect Human Height," *Nature* 467 (2010): 832.
15. T. A. Manolio, "Finding the Missing Heritability of Complex Diseases," *Nature* 461 (2009): 747–53; B. Maher, "The Case of the Missing Heritability," *Nature* 456 (2009): 18–21.
16. T. M. Teslovich et al., "Biological, Clinical and Population Relevance of 95 Loci for Blood Lipids," *Nature* 466 (2010): 707–13; A. R. Shuldiner and T. L. Pollin, "Variations in Blood Lipids," *Nature* 466 (2010): 703–4.
17. S. Menzel et al., "A QTL Influencing F Cell Production Maps to a Gene Encoding a Zinc-Finger Protein on Chromosome 2p15," *Nature Genetics* 39 (2007): 1197–99.
18. C. S. Ku et al., "The Pursuit of Genome-Wide Association Studies: Where Are We Now?" *Journal of Human Genetics* 55 (2010): 195–206. For a catalogue of GWA studies, see http://www.genome.gov/gwastudies/. Accessed April 24, 2010.
19. http://seer.cancer.gov/statfacts/html/all.html. Accessed April 24, 2010.
20. M. Serrano, "A Lower Bar for Senescence," *Nature* 464 (2010): 363–64.
21. Ku et al., "Pursuit of Genome-Wide Association Studies."
22. L. Ding et al., "Genome Remodelling in a Basal-like Breast Cancer Metastasis and Xenograft," *Nature* 464 (2010): 999–1005; J. Gray, "Genomics of Metastasis," *Nature* 464 (2010): 989–90.
23. Pearson, "Human Genetics."
24. Ibid.
25. Where I used to work, as it happens.
26. D. B. Kohn and F. Candotti, "Gene Therapy Fulfilling Its Promise," *New England Journal of Medicine* 360 (2009): 518–21.
27. A. Aiuti et al., "Gene Therapy for Immunodeficiency Due to Adenosine Deaminase Deficiency," *New England Journal of Medicine* 360 (2009): 447–58.

Chapter 10

1. L. O. Bygren et al., "Longevity Determined by Ancestors' Overnutrition during Their Slow Growth Period," *Acta Biotheoretica* 49 (2001): 53–59; M. E. Pembrey et al., "Sex-Specific, Male-Line Transgenerational Responses in Humans," *European Journal of Human Genetics* 14 (2006): 159–66.
2. G. Kaati et al., "Cardiovascular and Diabetes Mortality Determined by Nutri-

tion during Parents' and Grandparents' Slow Growth Period," *European Journal of Human Genetics* 10 (2002): 682–88.

3. M. E. Pembrey, "Time to Take Epigenetic Inheritance Seriously," *European Journal of Human Genetics* 10 (2002): 669–71.
4. Pembrey et al., "Sex-Specific, Male-line Transgenerational Responses in Humans."
5. Ibid.
6. L. H. Lumey et al., "Cohort Profile: The Dutch Hunger Winter Families Study," *International Journal of Epidemiology* 36 (2007): 1196–204.
7. R. C. Painter et al., "Transgenerational Effects of Prenatal Exposure to the Dutch Famine on Neonatal Adiposity and Health in Later Life," *BJOG: An International Journal of Obstetrics and Gynaecology* 115 (2008): 1243–49.
8. C. H. Waddington, "Preliminary Notes on the Development of the Wings in Normal and Mutant Strains of *Drosophila*," *Proceedings of the National Academy of Sciences USA* 25 (1939): 299–307.
9. A. Petronis, "Epigenetics as a Unifying Principle in the Aetiology of Complex Traits and Diseases," *Nature* 465 (2010): 721–27.
10. If you have a hard time remembering the difference between a eukaryote (cell that has a nucleus) and a prokaryote (cell with no nucleus), then just remember that "You" have a nucleus in your cells so you must be a "You-karyote." Being a person with a pictorial rather than abstract mind, I have always found these silly tricks quite useful, especially in exams. At school I once learned the whole periodic table of elements by making up a complete story, of which I can still remember the first line: Horrible Henry Lit a Beryllium Bomb Containing Nice Orange Flavored Nectar. Think about it.
11. A bit more jargon that you might find useful in following the scientific literature: "Euchromatin" refers to relaxed transcriptionally active DNA; "Heterochromatin" refers to condensed, more tightly packed DNA, which is less transcriptionally active. If you'd like a wacky YouTube-type pictorial description, then take a look at http://www.youtube.com/watch?v=Du_EJeCgjao. More serious attempts appear at http://www.youtube.com/watch?v=4b-NSWm24BA&NR=1 and http://www.youtube.com/watch?v=eYrQ0EhVCYA. All accessed May 21, 2010.
12. For example, see R. Margueron and D. Reinberg, "Chromatin Structure and the Inheritance of Epigenetic Information," *Nature Reviews Genetics* 11 (2010): 285–96.
13. C. A. Cooney et al., "Maternal Methyl Supplements in Mice Affect Epigenetic Variation and DNA Methylation of Offspring," *Journal of Nutrition* 132 (2002): 2393S–400S.
14. Ibid.; R. A. Waterland and R. L. Jirtle, "Transposable Elements: Targets for Early Nutritional Effects on Epigenetic Gene Regulation," *Molecular and Cellular Biology* 23 (2003): 5293–300.
15. K. Takahashi and S. Yamanaka, "Induction of Pluripotent Stem Cells from Mouse Embryonic and Adult Fibroblast Cultures by Defined Factors," *Cell* 126 (2006): 663–76; K. Hochedlinger and K. Plath, "Epigenetic Reprogramming and Induced Pluripotency," *Development* 136 (2009): 509–23; N. Maherali et al., "A High-Efficiency System for the Generation and Study of Human Induced Pluripotent Stem Cells," *Cell Stem Cell* 3 (2008): 340–45.

16. In practice, stem cells can be chosen that encode a different fur color from that encoded by the host blastocyst so that mice can eventually be bred according to their coat color. Those that have the "donor stem cell" coat color are the mice derived from the donor stem cells.
17. M. Stadtfeld, "Aberrant Silencing of Imprinted Genes on Chromosome 12qF1 in Mouse Induced Pluripotent Stem Cells," *Nature* 465 (2010): 175–83.
18. R. A. Brink, "A Genetic Change Associated with the R Locus in Maize Which Is Directed and Potentially Reversible," *Genetics* 41 (1956): 872–90.
19. A. A. Agrawal, "Transgenerational Induction of Defences in Animals and Plants," *Nature* 401 (1999): 60–63.
20. N. A. Youngson and E. Whitelaw, "Transgenerational Epigenetic Effects," *Annual Review of Genomics and Human Genetics* 9 (2008): 233–57.
21. F. M. Fraga et al., "Epigenetic Differences Arise during the Lifetime of Monozygotic Twins," *Proceedings of the National Academy of Sciences USA* 102 (2005): 10604–9.
22. K. Hayakawa et al., "Intrapair Differences of Physical Aging and Longevity in Identical Twins," *Acta Geneticae Medicae et Gemellologiae (Roma)* 41 (1992): 177–85.
23. Sibs or dizygotic twins do not share 50 percent of their genes in common, as is often mistakenly thought, but an average 50 percent of the amount of segregating DNA sequence variation that comes about due to random recombination during meiosis.
24. M. McGue et al., "Longevity Is Moderately Heritable in a Sample of Danish Twins Born 1870–1880," *Journal of Gerontology* 48 (1993): B237–B244.
25. B. Ljungquist et al., "The Effect of Genetic Factors for Longevity: A Comparison of Identical and Fraternal Twins in the Swedish Twin Registry," *Journals of Gerontology. Series A, Biological Sciences and Medical Sciences* 53 (1998): M441–M446.
26. Many of the world's early onset cases of Alzheimer's disease are found in and around the city of Medellín in Colombia. Members of twenty-five extended families with five thousand family members contain the bulk of these cases.
27. D. Mastroeni et al., "Epigenetic Differences in Cortical Neurons from a Pair of Monozygotic Twins Discordant for Alzheimer's Disease," *PLOS One* 4 (2009): e6617.
28. E. Korzus, "Manipulating the Brain with Epigenetics," *Nature Neuroscience* 13 (2010): 405–6.
29. C. Dulac, "Brain Function and Chromatin Plasticity," *Nature* 465 (2010): 728–35.
30. A. Prüss-Üstün and C. Corvalán, *Preventing Disease through Healthy Environments: Towards an Estimate of the Environmental Burden of Disease* (Geneva: World Health Organization, 2006).
31. Department of Health and Human Services, Centers for Disease Control and Prevention, 2005.
32. V. Bollati and A. Baccarelli, "Environmental Epigenetics," *Heredity* 105 (2010): 105.
33. Ibid.; M. P. Hitchins et al., "Inheritance of a Cancer-Associated MLH1 Germline Epimutation," *New England Journal of Medicine* 356 (2007): 697–705; R.

G. Gosden and A. P. Feinberg, "Genetics and Epigenetics: Nature's Pen-and-Pencil Set," *New England Journal of Medicine* 356 (2007): 731–33.

34. A. J. Wood and R. J. Oakey, "Genomic Imprinting in Mammals: Emerging Themes and Established Theories," *PLoS Genetics* 2 (2006): 1677–85.
35. For example, see ibid.
36. N. Krementsov, "Darwinism, Marxism, and Genetics in the Soviet Union," in *Biology and Ideology—From Descartes to Dawkins*, ed. Denis R. Alexander and Ronald N. Numbers (Chicago: University of Chicago Press, 2010), 215–46.
37. E. Pennisi, "The Case of the Midwife Toad: Fraud or Epigenetics?" *Science* 325 (2009): 1194–95; Arthur Koestler, *The Case of the Midwife Toad* (London: Hutchinson, 1971). The name "midwife" comes from the fact that the male carries a string of fertilized eggs around its legs.
38. A. Vargas, "Did Paul Kammerer Discover Epigenetic Inheritance? A Modern Look at the Controversial Midwife Toad Experiments," *Journal of Experimental Zoology Part B: Molecular and Developmental Evolution* 312B (2009): 667–78.
39. "Time for the Epigenome," *Nature* 463 (2010): 587.

Chapter 11

1. There is usually some rationale for naming these enzymes. In this case *Hind*III is an enzyme purified from *Haemophilus influenzae*. The naming of this bacterium in turn has an interesting history. It was first isolated in 1892 during an influenza epidemic and was thought at the time to be the cause of the influenza. Not until 1933 was influenza shown to be caused by viral rather than by bacterial infection, but the name has stuck (which is why, by the way, there is no point in taking antibiotics if you have the flu as they don't kill viruses). *Haemophilus influenzae* can cause other nasty diseases, including pneumonia and meningitis.
2. D. G. Gibson et al., "Creation of a Bacterial Cell Controlled by a Chemically Synthesized Genome," *Science* 329 (2010): 52–56.
3. Further details of approved therapeutic reagents used in medicine are available at www.biopharma.com/.
4. N. Cartier et al., "Hematopoietic Stem Cell Gene Therapy with a Lentiviral Vector in X-Linked Adrenoleukodystrophy," *Science* 326 (2009): 818–23.
5. D. H. Kim and J. J. Rossi, "Strategies for Silencing Human Disease Using RNA Interference," *Nature Reviews Genetics* 8 (2007): 173–84.
6. R. Zhao et al., "An Oncogenic Tyrosine Kinase Inhibits DNA Repair and DNA Damage-Induced Bcl-x_L Deamidation in T Cell Transformation," *Cancer Cell* 5 (2004): 37–49; R. Zhao et al., "DNA Damage-Induced Bcl-x_L Deamidation Is Mediated by NHE–1 Antiport Regulated Intracellular pH," *PLoS Biology* 5 (2007)doi:10.1371/journal.pbio.0050001; R. Zhao et al., "Inhibition of the Bcl-x_L Deamidation Pathway in Myeloproliferative Disorders," *New England Journal of Medicine* 359 (2008): 2778–89.
7. E. Cozzi and D. J. G. White, "The Generation of Transgenic Pigs as Potential Organ Donors for Humans," *Nature Medicine* 1 (1995): 964–66.
8. A. Pineyro-Nelson et al., "Transgenes in Mexican Maize: Molecular Evidence and Methodological Considerations for GMO Detection in Landrace Popula-

tions," *Molecular Ecology* 18 (2009): 750–61.

9. I. Potrykus, "Regulation Must Be Revolutionized," *Nature* 466 (2010): 561.
10. G. B. Kumar et al., "Expression of Hepatitis B Surface Antigen in Transgenic Banana Plants," *Planta* 222 (2005): 484–93.
11. H. Daniell, "Production of Biopharmaceuticals and Vaccines in Plants Via the Chloroplast Genome," *Biotechnology Journal* 1 (2006): 1071–79.
12. "The Growing Problem," *Nature* 466 (2010): 546–47.
13. R. Mittler and E. Blumwald, "Genetic Engineering for Modern Agriculture: Challenges and Perspectives," *Annual Reviews of Plant Biology* 61 (2010): 443–62.
14. V. Gewin, "An Underground Revolution," *Nature* 466 (2010): 552–53.
15. www.ers.usda.gov/Data/BiotechCrops/. Accessed July 9, 2010.
16. www.isaaa.org/resources/publications/briefs/39/highlights/default.html. Accessed July 9, 2010.
17. J. Tollefson, "The Global Farm," *Nature* 466 (2010): 554–55.
18. R. Paarlberg, *Starved for Science: How Biotechnology Is Being Kept Out of Africa* (Cambridge, MA: Harvard University Press, 2009).

Chapter 12

1. Denis R. Alexander, *Beyond Science* (Oxford: Lion Publishing, 1972).
2. Richard Dawkins, "The Ultraviolet Garden," Royal Institution Christmas Lecture No. 4, 1991.
3. www.nationmaster.com/graph/cri_pri_fem-crime-prisoners-female&int=-1. Accessed July 18, 2010.
4. Quoted in S. Rose, "Reflections on Reductionism," *Trends in Biochemical Sciences* 13 (1988): 160–62.
5. Francis H. Crick, *The Astonishing Hypothesis* (New York: Simon & Schuster, 1994), 3 (italics added).
6. Genesis 1:26–27; 5:1–2.
7. From Ibn Khaldun, *The Mukaddimah: An Introduction to History*, trans. Franz Rosenthal (Princeton: Princeton University Press, 1967).
8. My comments here on the compatibility of evolution with theism are necessarily brief, and I realize will not satisfy all the queries that might be raised. For a much fuller discussion, see Denis R. Alexander, *Creation or Evolution: Do We Have to Choose?* (Oxford: Monarch, 2008). A less theological treatment will be found in *Science, Evolution, and Creationism* (Washington, DC: National Academies Press, 2008).
9. Some of the books that I have found helpful in this context are the following: John Hick, *Evil and the God of Love* (London: MacMillan, 1985); W. B. Drees, ed., *Is Nature Ever Evil?* (New York: Routledge, 2003); M. M. Adams and R. M. Adams, *The Problem of Evil* (Oxford: Oxford University Press, 1990); M. M. Adams, *Horrendous Evils and the Goodness of God* (New York: Cornell University Press, 1999); J. V. Taylor, *The Christlike God* (London: SCM Press, 1992); M. S. Whorton, *Peril in Paradise* (Milton Keynes, England: Authentic, 2005); A. Elphinstone, *Freedom, Suffering and Love* (London: SCM Press, 1992); A. Farrer, *Love Almighty and Ills Unlimited* (London: Collins, 1962);

R. Swinburne, *Providence and the Problem of Evil* (Oxford: Oxford University Press, 1998); H. Blocher, *Evil and the Cross* (Grand Rapids: Kregel, 1994); B. Hebblethwaite, *Evil, Suffering and Religion* (London: SPCK, 2000); M. Larrimore, ed., *The Problem of Evil: A Reader* (Oxford: Blackwell, 2001); M. Murray, *Nature Red in Tooth and Claw: Theism and the Problem of Animal Suffering* (Oxford: Oxford University Press, 2008).

10. Revelation 21:4.
11. I am grateful to Professor Michael Murray for this parable, slightly adapted for my present purposes.
12. Michael Crichton, *Next* (New York: HarperCollins, 2006), 119.

Index